Omotola Olanrewaju

Modeling and Simulation of Fluid Catalytic Cracking (FCC) Riser

Omotola Olanrewaju

Modeling and Simulation of Fluid Catalytic Cracking (FCC) Riser

Mass transfer resistance, Coking and Products yield

LAP LAMBERT Academic Publishing

Impressum / Imprint
Bibliografische Information der Deutschen Nationalbibliothek: Die Deutsche Nationalbibliothek verzeichnet diese Publikation in der Deutschen Nationalbibliografie; detaillierte bibliografische Daten sind im Internet über http://dnb.d-nb.de abrufbar.

Alle in diesem Buch genannten Marken und Produktnamen unterliegen warenzeichen-, marken- oder patentrechtlichem Schutz bzw. sind Warenzeichen oder eingetragene Warenzeichen der jeweiligen Inhaber. Die Wiedergabe von Marken, Produktnamen, Gebrauchsnamen, Handelsnamen, Warenbezeichnungen u.s.w. in diesem Werk berechtigt auch ohne besondere Kennzeichnung nicht zu der Annahme, dass solche Namen im Sinne der Warenzeichen- und Markenschutzgesetzgebung als frei zu betrachten wären und daher von jedermann benutzt werden dürften.

Bibliographic information published by the Deutsche Nationalbibliothek: The Deutsche Nationalbibliothek lists this publication in the Deutsche Nationalbibliografie; detailed bibliographic data are available in the Internet at http://dnb.d-nb.de.

Any brand names and product names mentioned in this book are subject to trademark, brand or patent protection and are trademarks or registered trademarks of their respective holders. The use of brand names, product names, common names, trade names, product descriptions etc. even without a particular marking in this work is in no way to be construed to mean that such names may be regarded as unrestricted in respect of trademark and brand protection legislation and could thus be used by anyone.

Coverbild / Cover image: www.ingimage.com

Verlag / Publisher:
LAP LAMBERT Academic Publishing
ist ein Imprint der / is a trademark of
OmniScriptum GmbH & Co. KG
Heinrich-Böcking-Str. 6-8, 66121 Saarbrücken, Deutschland / Germany
Email: info@lap-publishing.com

Herstellung: siehe letzte Seite /
Printed at: see last page
ISBN: 978-3-659-71353-8

Copyright © 2015 OmniScriptum GmbH & Co. KG
Alle Rechte vorbehalten. / All rights reserved. Saarbrücken 2015

Acknowledgements

My heart-felt gratitude goes to the Executive Vice chancellor/CE, National Agency for Science and Engineering Infrastructure (NASENI), Prof. M. S. Haruna. I appreciate the Director Engineering Infrastructure, NASENI, Dr. M. Dauda. My gratitude also goes to the immediate past DG, NASENI, Prof. O. O. Adewoye (RIP). I am grateful to Dr. Steve Momoh for assistance rendered. I appreciate the entire management staff, NASENI for the permission and support accorded me by the Agency without which this work would not have been possible.

I wish to express my gratitude to my supervisors, Dr. P. C. Okonkwo and Prof. B. O. Aderemi for patiently overseeing this work from the beginning to the end. My heart-felt appreciation goes to all lecturers and staff of the Department of Chemical Engineering, ABU, Zaria, for all assistance rendered.

My appreciation also goes to my parents, Mr. and Mrs. Z. O. Osatuyi and my siblings: sister Dupe, brother Femi, Taiye, Kehinde (RIP) and Isaac for the endurance, patience and understanding exuded while I was engrossed by the work.

My sincere appreciation and unreserved love goes to my treasured celebrity *Angel* and wife, Olanrewaju Oluwakemi Esther (Sonia) for imparting my life during my studies in ABU, Zaria, and always being there for me as my *heartbeat*.

I appreciate Elder and Mrs. Femi Adeyele and family, Elder and Mrs. Julius Adeyele and family for the peculiar love shown to me.

I wish also to appreciate Prof. and Mrs. B. A. Omotara, Engr. and Mrs. Folorunsho Amodu, Engr. and Mrs. Oladimeji and Daddy Dele Anifowose and family for the roles played in my education. I appreciate Prof. Soboyejo, Dr. Omololu and Dr. Muktar, my academic mentors. I am extremely grateful to Bro. Yoni and aunty Alice and family for hospitality and always being there for me on the Campus. I appreciate Bro. Silas and Nonso Onyenanu; friends indeed that stuck closer than a brother. I am also grateful to Kehinde Salihu, Engr. Auwal, Mamza and Tumba for the assistance given to me in my time of dire need. I appreciate all my course mates. I am extremely grateful to my superiors and subordinates at work (Dr. Yusuf (RIP), Engr. Wale, Engr. Olasupo, Engr. Ajani, Engr. Asaolu, Mercy, Abayomi, Emma, Soji and Olowo, space will not permit me to mention you all). I am also grateful to Rev. Alex and my elders in the Lord.

Finally, I acknowledge with all reverence an Awesome *Personality*, Christ Jesus, for grace to be alive to pass through school again and for seeing *me* through in ABU, Zaria.

Dedication
I humbly dedicate this work to my Helper and Inspirer – *The Holy Ghost.*

Table of Contents

	Page
Acknowledgements	1
Dedication	2
Table of Contents	3
List of Figures	5
List of Tables	7
Nomenclature	8
1.0 INTRODUCTION	11
1.1 Preamble	11
1.2 Problem Statement	12
1.3 Aim	13
1.4 Objectives	13
1.5 Scope	13
1.6 Justification	13
2.0 LITERATURE REVIEW	14
2.1 Background	14
2.2 History of Fluid Catalytic Cracking (FCC)	14
2.3 Feedstock for Catalytic Cracking	15
2.4 Fluid Catalytic Cracking (FCC) Process Description	15
2.4.1 The feed preheat system	17
2.4.2 Reactor	17
2.4.3 Regenerator	18
2.5 Operational Variables of FCCU	19
2.5.1 Catalyst variables	19
2.5.2 Process variables	20
2.6 FCC Reactions and Lumping	20
2.6.1 The three-lump model	21
2.6.2 The four-lump model	22
2.6.3 Modeling of FCCU riser	23
2.6.4 Order of cracking reactions	29
2.6.5 Riser hydrodynamics	29
2.6.6 Catalyst deactivation	30

2.7	Coking of FCC Catalysts	31
2.8	Coking Models	32
2.8.1	Time-on-stream decay model	32
2.8.2	Decay model based on coke content	33
2.9	Kaduna Refinery and Petrochemicals Company (KRPC) Ltd	34
2.9.1	Catalytic cracking section	34
2.9.2	Cracking riser operating temperature	34
2.10	Riser Model Development	36
2.11	COMSOL Multiphysics and MATLAB	36
2.11.1	Unique features of COMSOL Multiphysics	37
2.11.2	Reaction Engineering Laboratory (REL)	37
2.11.3	Unique features of MATLAB	37
2.12	Derivation of Model Rate Equation	38
2.12.1	Effective diffusivity	39
2.12.2	Modeling of FCC reactions using the five-lump model	42
2.12.3	Development of two – dimensional pseudo-homogeneous riser model	44
3.0	MATERIALS AND METHODOLOGY	48
3.1	Modeling Strategy	48
3.1.1	FCC reaction modeling	48
3.1.2	Modeling of riser	49
3.1.3	Finite difference discretization of the governing equations	50
3.2	Riser Kinetic Data	53
4.0	RESULTS AND DISCUSSION	55
4.1	Reaction Model Results (COMSOL Multiphysics)	55
4.2	Reactor Model Results (MATLAB)	60
4.2.1	Validation of model results	64
4.3	Simulation of Coking	65
5.0	CONCLUSIONS AND RECOMMENDATIONS	67
5.1	Conclusions	67
5.2	Recommendations	68
REFERENCES		69

List of Figures

Figure No.	Title	Page No.
Figure 2.1	Universal Oil Product (UOP) type Fluid Catalytic Cracking Unit	16
Figure 2.2	Typical riser of a Fluid Catalytic Cracking Unit (FCCU)	17
Figure 2.3	Rate constants of the three-lump model	21
Figure 2.4	Rate constants of the four-lump model	22
Figure 2.5	Five-lump model for catalytic cracking of Vacuum Gas Oil (VGO)	24
Figure 2.6	Fluid Catalytic Cracking (FCC) species concentration profiles	25
Figure 2.7	Catalyst deactivation and coke concentration profiles of Fluid Catalytic Cracking	26
Figure 2.8	Gasoline yield and gas oil conversion in Fluid Catalytic Cracking (FCC)	27
Figure 2.9	Products yield in Fluid Catalytic Cracking (FCC)	27
Figure 2.10	Temperature profile in Fluid Catalytic Cracking Unit (FCCU) riser	28
Figure 2.11	Axial velocity profiles of gas and solid phases in the riser	30
Figure 2.12	Concentration and temperature gradients in a porous catalyst	39
Figure 2.13(a)	2D riser	44
Figure 2.13(b)	Control volume for riser model	44
Figure 3.1	Finite difference grid	50
Figure 3.2	Discretization of second order differentials	50
Figure 3.3	Sequence of work for the simulation of Fluid Catalytic Cracking (FCC) reactions	52
Figure 3.4	Computational flow diagram for riser model	53
Figure 4.1	Effect of mass transfer resistance on Fluid Catalytic Cracking (FCC) reactions (no coking)	55
Figure 4.2	Effects of mass transfer resistance and catalyst coking on selectivity and yield of Fluid Catalytic Cracking (FCC)	56
Figure 4.3	Comparison of model predictions and predictions from literature	57
Figure 4.4	Coke concentration and catalyst activity decline in Fluid Catalytic Cracking (FCC) of Vacuum Gas Oil (VGO)	60

Figure 4.5	Products yield along riser height ---61
Figure 4.6	Conversion of Vacuum Gas Oil (VGO) along riser height ---62
Figure 4.7	Temperature drop along riser height ---62
Figure 4.8	Sections of Fluid Catalytic Cracking (FCC) riser ---63
Figure 4.9	Model result for simulation of coking in Fluid Catalytic Cracking (FCC) riser ---65
Figure 4.10	Variation of Catalyst-Oil-Ratio (COR) and gasoline yield with reaction temperature ---66

List of Tables

Table No.	Title	Page No.
Table 2.1	History of catalytic cracking	14
Table 2.2	Comparison of thermal cracking and catalytic cracking	15
Table 2.3	Previous related works	23
Table 2.4	Kinetic constants for five-lump model	24
Table 2.5	Kaduna Refinery and Petrochemicals Company (KRPC) feedstock	34
Table 2.6	Feedstock and product design data for Kaduna Refinery and Petrochemicals Company (KRPC) Riser Catalytic Cracking Unit (RCCU)	35
Table 3.1	Enthalpies of cracking	53
Table 3.2	Molecular weights and heat capacities	53
Table 3.3	Gas oil properties	54
Table 3.4	Riser model parameters	54
Table 4.1	Kaduna Refinery and Petrochemicals (KRPC) Fluid Catalytic Cracking Riser (FCCR) flow rates (Abridged version, May, 2013)	59
Table 4.2	Kaduna Refinery and Petrochemicals (KRPC) Fluid Catalytic Cracking Riser (FCCR) design data	59
Table 4.3	Validation of model results with plant data	64
Table 4.4	Validation of model results with results from literature	64

Nomenclature

A_c	Cross-sectional area (m²)
a	Catalyst activity for non-coking reactions
b	Catalyst activity function for coking reactions
c_i	Species concentration (weight fraction)
c_p	Specific heat capacity (J/kg-K)
d_{AB}	Collision diameter (m)
D_{AB}	Molecular diffusivity (m/s²)
D_e	Effective diffusivity (m/s²)
D_k	Knudsen diffusivity (m/s²)
D_p	Particle diameter (m)
D^*	Overall diffusivity (m/s²)
F_r	Froud number
F_i	Flow rate of species i (kg/s)
G_s	Catalyst mass flux (kg/m².s)
H	Reactor height (m)
ΔH_{Ri}	Enthalpy of cracking of species i (kJ/kg)
ΔH_{vap}	Enthalpy of vaporization (kJ/kg)
k	Reaction rate constant (s⁻¹)
k_g	Mass transfer coefficient (m/s)
k_r, k_z	Effective thermal conductivity (W/m.K)
M_i	Molecular weight species i (kg/kmol)
m	Node number in the horizontal direction
n	Node number in the vertical direction
N_A	Molar flux (kmol/m².s)
N_{Re}	Particle Reynolds number
N_{sc}	Schmidt number
N_{sh}	Sherwood number
N_r	Number of divisions in radial direction
N_z	Number of divisions in axial direction
P	Pressure (atm)
R	Radius (m)
R_g	Universal gas constant (J/K-mol)
Δr	Radial spatial interval (m)
r_e	Average pore radius (m)

q	Volumetric flow rate (m³/s)
r_i	Species reaction rate (kg species (kg catalyst)$^{-1}$s^{-1})
t	Time (s)
T	Temperature (K)
u	Superficial velocity (m/s)
V	Reactor volume (m³)
v_{ij}	Stoichiometric coefficient
v_p	Average particle velocity (m/s)
X	Conversion
Δz	Axial spatial interval (m)

Greek letters

α'	Decay function rate constant
α	Normalized parameter
β	Normalized parameter
γ	Normalized parameter
δ	Decay function constant
ε	Porosity
η	Particle effectiveness factor
η_0	Particle overall effectiveness factor
λ	Normalized parameter
μ	Viscosity (Pa.s^{-1})
π	Pi
φ	Thiele modulus
ψ	Slip factor
ρ	Density (kg/m³)
σ	Normalized variable
τ	Tortuosity
Ω_D	Collision integral

CHAPTER 1
INTRODUCTION
1.1 Preamble

Fluid catalytic cracking is one of the most profitable processes in oil refineries. It upgrades heavy petroleum fractions to more valuable lighter products by cracking. It is the major producer of gasoline in refineries and as such it is sometimes referred to as the heart of the refinery (Fernandes *et al.*, 2003).

There are approximately four hundred (400) fluid catalytic crackers operating worldwide with a total processing capacity of over 45,000m^3/day (12 million barrels per day) (Gupta, 2006). Oil companies such as Exxon, Shell and TOTAL have their own designs. However most of the current operating units have been designed or revamped by Universal Oil Products (UOP), M. W. Kellog and Stone and Webster (Gupta, 2006). Fluid Catalytic Cracking (FCC) has been used for more than 60 years in order to convert straight-run atmospheric gas oils, vacuum gas oils, atmospheric residues and heavy stocks recovered from other refinery operations into high octane gasoline, light fuel oils and olefin-rich light gases (Bessiris and Harismiadis, 2007).

In a refinery, crude oil is distilled in an atmospheric distillation unit to produce Liquefied Petroleum Gas (LPG), naphtha, kerosene and diesel oil. The residue from the atmospheric distillation unit is fed to a vacuum distillation unit where it is separated into vacuum gas oils and vacuum residue. The heavy vacuum gas oil, usually 25-30% of the total crude oil volume, is fed to the Fluid Catalytic Cracking Unit (FCCU) where it is converted into lighter products. Gupta (2006) reported that the heavy vacuum gas oil (VGO) has a boiling range of 621K (348^{0}C) to 838K (565^{0}C). Other feedstocks that are processed in FCC units include hydrotreated gas oils, cracked gas oils and deasphalted oils. The FCCU is often regarded as the "garbage disposal" of the refinery because of its ability to process this wide range of feeds into useful products. The typical process variables in FCC operation are temperature, catalyst-to-oil ratio and pressure. Of these variables, the process parameter that has the greatest influence on products yield and distribution is the reactor temperature. The reactor temperature is varied by varying the catalyst-to-oil ratio. Pressure however, is varied over a narrow range in FCC operation due to limited blower capacity (Kellogg, 1980).

In a typical FCC reactor, feed distributors inject the feed into the riser to contact it with the catalyst and initiate the reaction. A highly efficient stripper then separates the remaining hydrocarbon from the catalyst and the catalyst is sent to the regenerator to be reactivated. The reactivated catalyst flows back to the mixing chamber to be contacted again with the feed. The reactor zone usually consists of a short-contact riser and a termination device for quick separation of catalyst from the hydrocarbon products (Gupta, 2006).

Modeling of the reactor section of FCC unit has been an active area of research in industry and academia alike (McFarlane *et al.*, 1990). Examples of existing models of the riser are: the three-lump model of Weekman and Nace (1970), the four-lump model of Ali and Rohanni (1997), the four-lump model of Ahari *et al.* (2008), the five-lump model of Den Hollander *et al.* (2003). Under normal FCC conditions coke is the most important factor that affects catalyst activity. As catalytic reactions proceed, coke deposits on the catalyst surface. Coke covers the catalyst active sites leading to catalyst activity decay. This underscores the need for riser models that can adequately simulate catalyst coking in FCC reactions. It was observed that the three-lump model of Weekman and Nace could not predict coke yield whereas the other models accounted for coking in FCC reactions.

Existing riser models were also based on the assumptions of 1D plug flow and negligible mass transfer resistance. These assumptions have resulted in the under-prediction of the overcracking point of gasoline and the overprediction of the of the riser residence time in the existing models. In reality the riser is a 3D reactor. Simplifying the geometry to 1D is tantamount to predicting products yield along the axis of the reactor. However turbulent the flow in the riser may be, a 1D model cannot adequately represent the entire geometry of the reactor because it does not account for wall effects. Negligible mass transfer resistance on the other hand implies that the reacting species encounter no resistance in diffusing from the bulk gas phase to the active sites of the porous catalysts. This is against what is established in technical literature concerning heterogeneous catalytic reactions; concentration gradient always exists in heterogeneous catalysis. Models that are based on negligible mass transfer resistance predict shorter contact time for FCC reactions than obtains in reality.

1.2 Problem Statement

The assumption of 1D plug flow with negligible mass transfer resistance is responsible for the overprediction of the riser residence time and the underprediction of gasoline overcracking point in the existing models.

1.3 Aim

The aim of this work was to model and simulate coking and selectivity of Fluid Catalytic Cracking (FCC) using a 2D riser model with mass transfer resistance considered.

1.4 Objectives

The objectives of this work were as follows:
1. To model the cracking reactions occurring in the FCC unit riser using a five-lump reaction scheme.
2. To solve the model using COMSOL Multiphysics 4.0 and MATLAB.
3. To determine the effect of catalyst coking on FCC products yield and selectivity.
4. To determine the effect of variation of reaction temperature (or catalyst-to-oil ratio) on coking of the catalyst.

1.5 Scope

In this book, a five-lump reaction model was used to model the effects of mass transfer resistance and coking on FCC reactions (reaction modeling). The riser of an industrial scale FCC unit was simulated (reactor/geometry modeling). Product data from Kaduna Refinery and Petrochemicals Company (KRPC) Ltd and data from literature were used to validate the model. Finally, the effect of varying reactor temperature (or catalyst-to-oil ratio) on coking of FCC catalyst was investigated.

1.6 Justification

Though life pilot plant units provide good simulation for commercial FCC units, they are expensive, difficult and costly to operate. Moreover, riser simulators such as the Chemical Reactor Engineering Centre (CREC) riser simulator are not locally available for researchers in developing countries to easily access and generate riser experimental and model data.

CHAPTER 2
LITERATURE REVIEW
2.1 Background

Catalytic cracking involves contacting a feedstock (usually a gas oil fraction) with a catalyst under suitable conditions of temperature, pressure and residence time. A substantial part of the feedstock (>50%) is converted into gasoline and lower-boiling products in a single-pass operation. Fluid catalytic cracking has been used for more than 60 years to convert heavy petroleum fractions into lighter and more valuable components (Bessiris and Harismiadis, 2007). A modern FCC unit comprises of a riser, a stripper, a disengager, a regenerator, a main fractionator, catalyst transport lines (spent catalyst standpipe and regenerated catalyst standpipe) together with several other auxiliary units such as cyclone, air blower, expander, wet gas compressor, feed pre-heater, air heater, catalyst cooler, CO boiler, etc (Fernandes *et al.*, 2003). The conversion of heavy hydrocarbon fractions into lower molecular weight products takes place in the riser.

2.2 History of Fluid Catalytic Cracking (FCC)

Catalytic cracking is a conversion process that can be applied to a variety of feed stocks ranging from gas oil to heavy oil. It differs from thermal cracking in that it uses a catalyst that is not (in theory) consumed in the process. The original incentive to develop cracking processes arose from the need to increase gasoline supplies and to increase the octane number of gasoline while maintaining yield from high-boiling stocks. The purpose of cracking was wholly justified because it could virtually double the volume of gasoline from a barrel of crude oil (Speight, 2007).

Table 2.1 gives a summary of the history of catalytic cracking while Table 2.2 compares thermal cracking and catalytic cracking.

Table 2.1: History of catalytic cracking (Speight, 2007)

Year	Description
1915:	Batch reactor catalytic cracking to produce light distillates. Catalyst: aluminium chloride ($AlCl_3$)- a Lewis acid, electron acceptor. Alkane-electron (abstracted by aluminium chloride) to produce carbocation (C^+). Ionic chain reactions to crack long chains.
1936:	Houndry process; continuous feedstock flow with multiple fixed-bed reactors. Catalyst: clays, natural alumina/silica particles.
1942:	Thermoform catalytic cracking with moving-bed catalyst. Continuous feedstock flow.

Catalyst: synthetic alumina/silica particles; higher thermal efficiency by process integration.

1942: Fluid Catalytic Cracking (FCC) with fluidized-bed catalyst. Continuous feedstock flow.

1965: Promising new catalysts; synthetic alumina/silica and zeolite catalysts.

Table 2.2: Comparison of thermal cracking and catalytic cracking (Speight, 2007)

Thermal	Catalytic
1. No catalyst	Uses catalyst
2. Higher temperature	Lower temperature
3. Higher pressure	Lower pressure
4. Free-radical reaction mechanisms	Ionic reaction mechanisms
5. Moderate thermal efficiency	High thermal efficiency
6. No regeneration of catalyst needed	Good integration of cracking and regeneration.
7. Moderate yields of gasoline and other distillates	High yields of gasoline and other distillates
8. Low-to-moderate product selectivity.	High product selectivity.
9. Low octane number gasoline.	High octane number.
10. Low-to-moderate yield of C4 olefins	High yield of C4 olefins
11. Low-to-moderate yields of aromatics	High yields of aromatics

2.3 Feedstock for Catalytic Cracking

The feedstock for catalytic cracking can be:

1. Straight-run gas oil
2. Vacuum gas oil
3. Atmospheric residuum
4. Vacuum residuum

The pretreatment options for the feedstock for catalytic cracking units are:

1. Deasphalting to prevent excessive coking on catalyst surfaces.

2. Demetallation: removal of nickel, vanadium and iron to prevent catalyst deactivation.

3. Hydrotreating or mild hydrocracking to prevent excessive coking in the Fluid Catalytic Cracking Unit (FCCU).

2.4 Fluid Catalytic Cracking (FCC) Process Description

All modern FCC units consist of two major parts: the riser (where hydrocarbon feed contacts the catalyst to initiate cracking reaction) and the regenerator wherein the spent catalyst is reactivated by burning off the coke that is deposited on the catalyst. Figure 2.1 gives an illustration of a typical Fluid Catalytic Cracking Unit (FCCU) while Figure 2.2 gives an illustration of the riser.

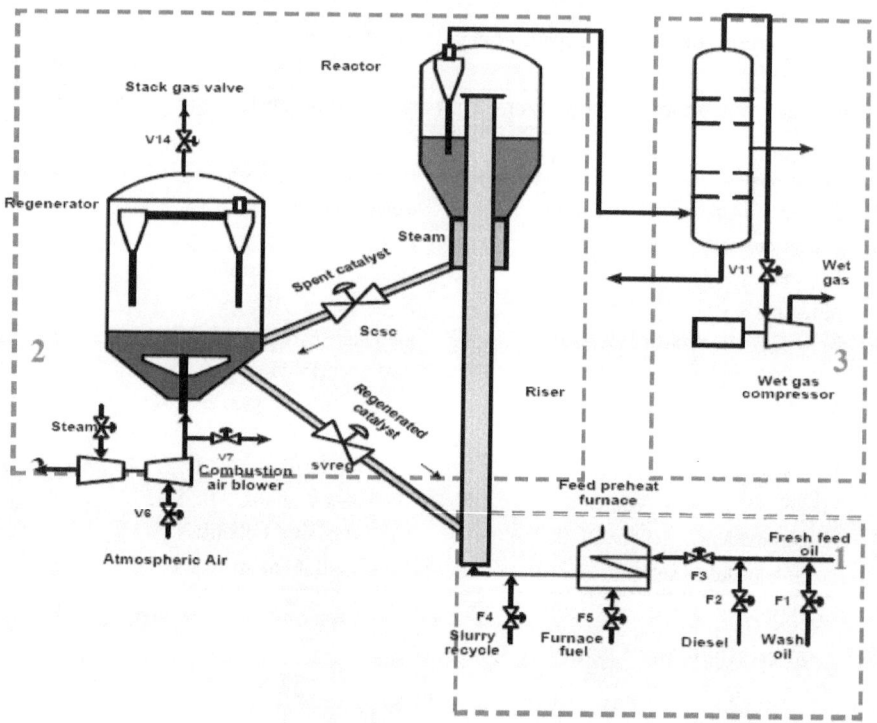

Figure 2.1: Universal Oil Product (UOP) type Fluid Catalytic Cracking Unit (Alsabei, 2011)

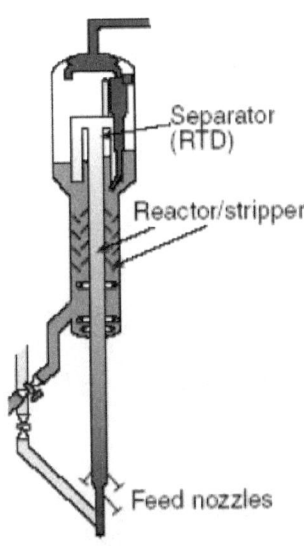

Figure 2.2: Typical riser of a Fluid Catalytic Cracking Unit (FCCU) (Fernandes et al., 2003)

2.4.1 The feed preheat system

In the refinery, gas oil from the vacuum distillation unit and supplemental feed stocks are combined and sent to a surge drum. The surge drum provides a constant flow of feed to the FCC unit's charge pumps. The drum is also used to separate any water or vapour that may be present in the feedstocks. The feed from the surge drum is heated to a temperature of 543K-648K (270^0C-375^0C) before it is fed to the FCCU reactor (Fernandes et al., 2003). The feed is preheated usually by exchange with light crude oil or bottom pumparounds. Preheating can also be done by using fired heaters.

2.4.2 Reactor

Almost all FCC reactors consist of a riser in which feed and catalyst are contacted for a very short time (less than 5seconds). Most of the cracking reactions occur in the riser before catalysts and products are separated in the reactor. Notwithstanding some undesired catalytic cracking reactions continue to occur in the reactor housing. The feed from the preheater enters the riser close to its base. It is therein contacted with hot regenerated catalyst from the regenerator. The ratio of catalyst to oil is reported to be in the range of 4:1 to 9:1 by weight (Gupta, 2006). Cracking reactions are endothermic. The heat required to heat and vaporize the feed as well as initiate the cracking is supplied by the heated catalyst from the regenerator. The feed is vaporized as soon as it contacts the catalyst and the reaction commences immediately after vaporization. The expansion of the vaporized feed drags the catalysts up the riser.

According to Gupta, (2006) riser dimensions are about ½ a meter to 2m in diameter and 20 to 35m in length. Ideally the riser is a plug flow reactor. In reality, the catalyst particles slip while rising with the vapour in the reactor and there is significant back mixing of particles. For the desired cracking reactions to occur, efficient contact of feed and catalyst is necessary. Atomization of feed is carried out by steam. This enhances the presence of the feed at the reactive acid site of the catalyst. The cracking reactions take place in the riser in two to four seconds when high-activity zeolite catalyst is used. Risers are designed to have an outlet vapour velocity of 15-20m/s and an average hydrocarbon residence time of about 2s at outlet conditions (Fernandes *et al.*, 2003).

The product vapours and the catalyst flow through a Riser Termination Device (RTD) at the end of the riser. This device separates the catalyst from the hydrocarbon vapour. Overcracking of products is avoided by carrying out the separation quickly. The separated catalyst is channeled to the catalyst stripper where hydrocarbon vapour adhering to it is stripped off by steam. The hydrocarbon vapour from the RTD enters the reactor vessel. The vessel provides disengagement space between the RTD and the cyclones.

Product vapours entering the reactor vessel from the RTD get mixed with steam and stripped hydrocarbon vapours and then flow through the reactor cyclones into the main fractionator.

2.4.3 Regenerator

The regenerator in the FCCU performs two main functions:
1. It restores the activity of the catalyst.
2. It supplies the heat needed to initiate the cracking reactions.

The spent catalyst entering the regenerator is reported to contain 0.25-0.5wt% coke (Gupta, 2006). Coke constitutes carbon, hydrogen and trace amounts of sulfur and nitrogen. The coke deposited on catalyst surface does not have a specific molecular structure. The commonly used formula for coke is CHn, where n is normally taken from 0.4 to 2.0 (Gupta, 2006).

Deactivated catalyst is regenerated by passing air through it in the regenerator. The air velocity and pressure is such that can maintain the catalyst in fluidized state. Oxygen in the air burns off the coke on the catalyst. It is this combustion that raises the temperature of the regenerated catalysts.

Regenerated catalyst enters the riser through the regenerated catalyst standpipe. The riser outlet temperature is controlled by a slide valve in the regenerated catalyst standpipe by controlling the quantity of hot catalyst entering the riser. The flue gases leaving the regenerator flow through the regenerator cyclones wherein entrained catalysts are separated and returned to the regenerator.

2.5 Operational Variables of FCCU

The major process variables in FCC are temperature, pressure, catalyst-to-feedstock ratio (ratio of the weight of catalyst entering the reactor per hour to the weight of feedstock charged per hour), and space velocity (weight or volume of feedstock charged per hour per weight or volume of catalyst in the reaction zone). Increased conversion can be obtained by applying higher temperature or higher pressure. Alternatively, lower space velocity and higher catalyst-to-feedstock ratio will contribute to increased conversion (Speight, 2007).

Single-stage cracking causes more reactive hydrocarbon to be cracked with a high conversion to gas and coke in the reaction time necessary for reasonable conversion of the more refractory hydrocarbons. However, in a two-stage process, gas and gasoline from short-reaction-time, high-temperature cracking operations are separated before the main cracking reactions take place in the second stage reactor (Speight, 2007).

The primary variables in the operation of FCC units for maximum unit conversion are categorized into two:
- Catalyst variables
- Process variables

2.5.1 Catalyst variables

The primary variables available to the design of FCC units for maximum unit conversion for a given feedstock quality are the catalyst variables. These variables include catalyst activity and catalyst design (which includes availability of cracking sites and the presence of carbon on regenerated catalyst).

Activity is increased by one or combination of:
1. Increased fresh catalyst addition rate
2. Increased fresh catalyst zeolite activity

Increasing the concentration of catalyst in the reactor often referred to as catalyst-to-oil ratio, will increase the availability of cracking for maximum conversion provided that the unit is not already operating at a catalyst circulation limit (Speight, 2007).

2.5.2 Process variables

The process variables include:
- Pressure
- Reaction/contact time
- Reaction temperature

Higher conversion and coke yields are thermodynamically favoured by higher pressure. However, pressure is usually varied over a very narrow range due to limited air blower horsepower.

An increase in reaction time available for cracking also increases conversion. Fresh feed rate, riser steam rate and recycle rate are the primary operating variables that affect reaction time for a given unit configuration. As a result of limited reactor size available for cracking, conversion varies inversely with the steam rates aforementioned. Increased reactor temperature increases feedstock conversion, primarily through a higher rate of reaction for the endothermic cracking reaction and also through increased catalyst-to-oil ratio (Speight, 2007). Reaction temperature is the variable most readily changed in FCC unit operation.

2.6 FCC Reactions and Lumping

The kinetic modeling of catalytic cracking has been traditionally based on using a lumping strategy (chemical species with similar behaviours are grouped together forming a smaller number of "pseudo" species).

Generally, there are two basic techniques in lumping the catalytic cracking of VGO (>633K). The first approach is to lump molecules in different distillation cuts (pseudo-species) and to consider chemical reactions between these lumps. The lumps usually constitute the feedstock and the final cracking products: gasoline (C_5-493K),

LCO (493-633K), light gases (C_3 and C_4) and coke. The second approach is to lump different products based on main chemical families such as paraffins, olefins, naphthenes and aromatics (Gupta, 2006).

2.6.1 The three-lump model

Weekman and Nace (1970) in their pioneer work developed a simple kinetic scheme (the three-lump scheme) for the kinetic modeling of cracking reactions in the riser. The authors divided the charge stock and products into three components: the initial feedstock, the gasoline and the remaining C_4s (dry gas and coke). This division oversimplified the scheme. The model predicted the conversion of gas oil (the feedstock) and gasoline yield in isothermal condition in fixed, moving and fluid bed reactors. The three-lump model however, could not predict the yield of coke because coke was lumped with dry gas in the model.

The rate constants of the reactions in the three-lump model are shown in Figure 2.3.

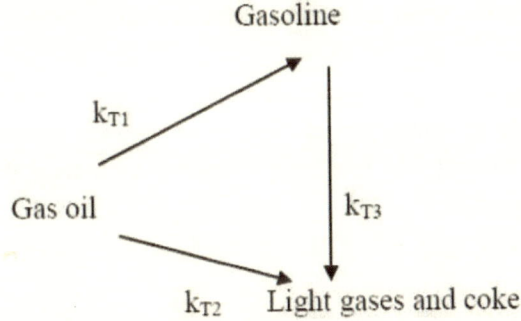

Figure 2.3: Rate constants of the three-lump model ("Lumping and Modeling", 1996)

The rate equations for the three-lump model as given in literature are:

i) Gas oil cracking:

$$r_{go} = k_{T0}\phi_1 C_{go}^2 \qquad 2.1$$

$$k_{T0} = -(k_{T1} + k_{T2})$$

ii) Gasoline formation:

$$r_g = k_{T1}\phi_1 C_{go}^2 - k_{T3}\phi_2 C_g \qquad 2.2$$

iii) Light gases and coke formation:

$$r_c = k_{T2}\phi_1 C_{go}^2 + k_{T3}\phi_2 C_g \qquad 2.3$$

In Equations 2.1 to 2.3, k_{T0}, k_{T1}, k_{T2} and k_{T3} represent the overall VGO cracking rate constant, the gasoline formation rate constant, the coke and gases formation rate constant and the gasoline overcracking rate constants respectively, ϕ represents catalyst activity.

The three-lump model paved way for the development of several other kinetic schemes. Such schemes included the four-lump model (Yen et al., 1988; Lee et al., 1989), five-lump model (Larocca et al., 1990), six-lump model (Coxon and Bischoff, 1987, Takatsuka et al., 1987), ten-lump model (Jacob et al., 1976), eleven-lump model (Mao et al., 1985; Sa et al., 1985; Zhu et al., 1985), twelve-lump model (Oliveira, 1987), thirteen-lump model (Sa et al., 1995) and nineteen-lump model (Pitault et al., 1994).

2.6.2 The four-lump model

Yen et al., (1987) introduced the four-lump model by splitting the light gases + coke lump into separate lumps (coke and light gases). The authors used a second order reaction for VGO cracking. The model is reported to predict very effectively the coke yield for VGO cracking in FCC pilot plants and commercial units. However, the work of the authors ended at predicting the yield of coke, thereby leaving room for more work to be done precisely on simulating the effect of variation of process parameters on the coking of FCC catalyst. The rate constants of the reactions in the four-lump model are shown in Figure 2.4.

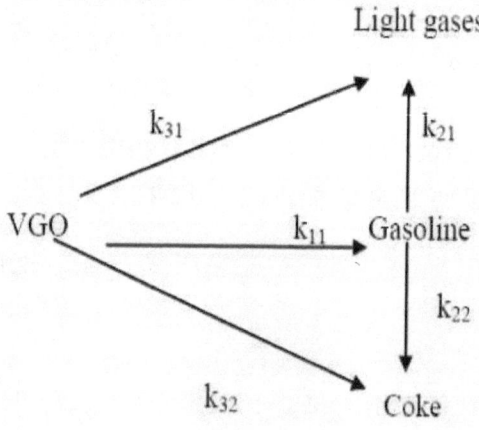

Figure 2.4: Rate constants of the four-lump model ("Lumping and Modeling", 1996)

The rates of gasoline formation and consumption were described as follows ("Lumping and Modeling", 1996):

i) Gas oil consumption rate:
$$r_{go} = k_{11}\phi C_{go}^2 + k_{31}\phi C_{go}^2 + k_{32}\phi C_{go}^2 \qquad 2.4$$

ii) Gasoline formation rate:
$$r_g = k_{11}\phi C_{go}^2 - \phi(k_{21} + k_{22})C_g \qquad 2.5$$

iii) Light gases formation rate:
$$r_{lg} = k_{31}\phi C_{go}^2 + k_{21}\phi C_g \qquad 2.6$$

iv) Coke formation rate:
$$r_c = k_{32}\phi C_{go}^2 + k_{22}\phi C_g \qquad 2.7$$

Juarez et al., (1997) extended the four-lump model to five lumps. They divided the gas lump into two different lumps, viz: dry gas and liquefied petroleum gas (LPG). LPG can be formed either directly from gas oil or as a secondary product from gasoline overcracking. Dry gas (H_2, C_1, C_2) can be formed either directly from gasoline and liquefied petroleum gas (LPG) cracking.

2.6.3 Modeling of FCCU riser

The modeling of the riser reactor of an FCCU is complex as a result of the presence of three phases (solid, liquid and vapour), involvement of physical and chemical rate steps and the strong interaction of the reactor with the regenerator. Considerable efforts have been made by various researchers in the modeling of FCCU riser. A summary of the main features of the available riser models is given in Table 2.3.

Table 2.3: Previous related works

	Den Hollander et al. (2003)	You & Zhu (2008)	Praveen and Shishir (2009)	Heydari et al. (2010)	Alsabei (2011)
Vaporization	Instantaneous	Instantaneous	Instantaneous	Instantaneous	Instantaneous
Temperature variation	Isothermal	Adiabatic	Adiabatic	Adiabatic	Adiabatic
Molar expansion	Not considered	-	Not considered	Considered	Considered
Axial catalyst Holdup	Single particle dynamics	Single particle dynamics	Single particle dynamics	-	-
Mass Transfer	Not considered	Not considered	Not considered	Not considered	Not considered

Kinetic model	Five-lump	Four-lump	Five-lump	Four-lump	Five-lump
Deactivation	Non-selective, based on time-on-stream, coke content of catalyst	Non-selective, based on catalyst coke concentration	Non-selective, based on time-on-stream of catalyst	Non-selective, based on time-on-stream of catalyst	Non-selective, based on catalyst coke concentration

Den Hollander *et al.* (2003) used a five-lump scheme (Figure 2.5) to model the concentration transients of the lumps in the model. Their model was able to predict the yield of coke but overpredicted gasoline yield while the overcracking point of gasoline was underpredicted as a result of negligence of mass transfer resistance. The profiles obtained by the authors are as shown in Figure 2.6 and Figure 2.7.

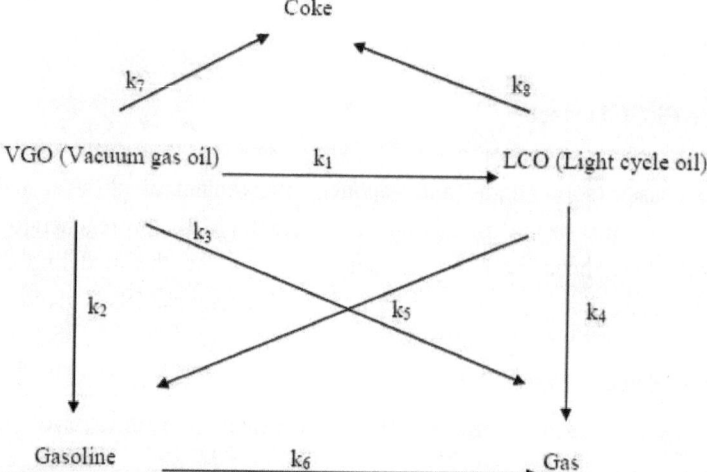

Figure 2.5: Five-lump model for catalytic cracking of Vacuum Gas Oil (VGO) (Den Hollander *et al.*, 2003)

Table 2.4 gives the literature values of the kinetic constants shown in the five-lump model above (Figure 2.5).

Table 2.4: Kinetic constants for five-lump model (Den Hollander *et al.*, 2003)

Reaction number	k (s^{-1})
1	1.90
2	7.50
3	1.50

4	0.00
5	1.00
6	0.30
7	0.21
8	0.50

The advantages of the five-lump model are as follows:

1) The model can predict catalyst coking (unlike the three-lump model which does not treat coke as a separate lump).

2) The model also captures the major species involved in FCC reactions (Gas oil, light cycle oil, gasoline, gas and coke).

3) The five-lump model is more flexible to analyze and to code/solve than the models with more lumps.

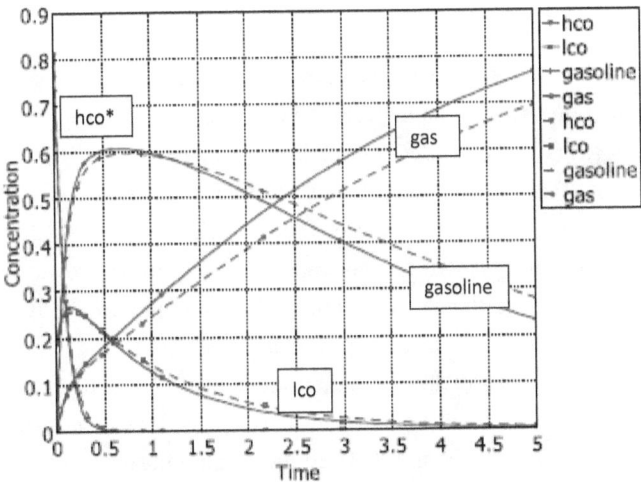

Figure 2.6: Fluid Catalytic Cracking (FCC) species concentration profiles (— : Concentration profiles without catalyst coking - - - : Concentration profiles when coking of catalyst occurs *hco: Heavy cycle oil) (Den Hollander et al., 2003)

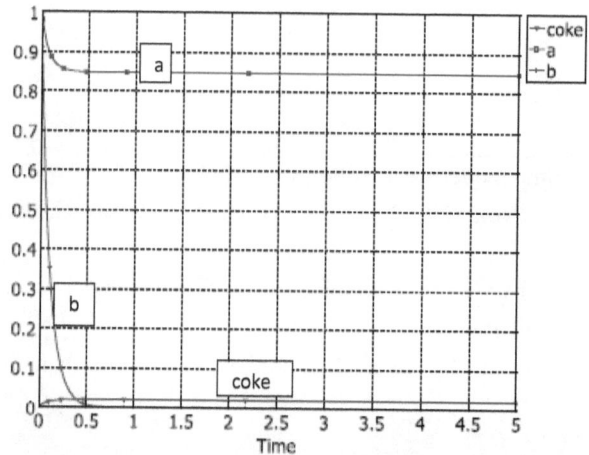

Figure 2.7: Catalyst deactivation and coke concentration profiles in Fluid Catalytic Cracking (a: Catalyst activity profile for non-coking reactions b: Catalyst activity profile for coking reactions) (Den Hollander et al., 2003)

You and Zhu (2008) used a 1D, four-lump model to predict the products yield of FCCU riser, the average gas density, solid and gas phase velocity profiles all as a function of the riser height. They predicted a gasoline yield of 45%. However, non-incorporation of mass transfer resistance in their model oversimplified the model. Praveen and Shishir (2009) used a 1D, five-lump model in their investigation. They predicted the temperature drop along the riser but their model was also oversimplified because of negligence of dispersion and assumption of 1D plug flow. A four-lump, 1D scheme was also used by Heydari et al. (2010) to model an industrial riser. They predicted the yield of gasoline along the riser under varying conditions of temperature and catalyst-to-oil ratio. Their model was also oversimplified and they did not simulate catalyst coke content. A five-lump reaction scheme was used by Alsabei (2011). The author also based his investigation on negligible dispersion which contradicts the basic principles of solid catalysis for porous catalysts such as the FCC catalyst.

Ahari et al. (2008) used a four-lump scheme to model an industrial scale riser reactor. 1D plug flow without mass transfer resistance was assumed in their work and the authors predicted a gasoline yield of 45%. The products yield obtained by the authors is given in Figure 2.8.

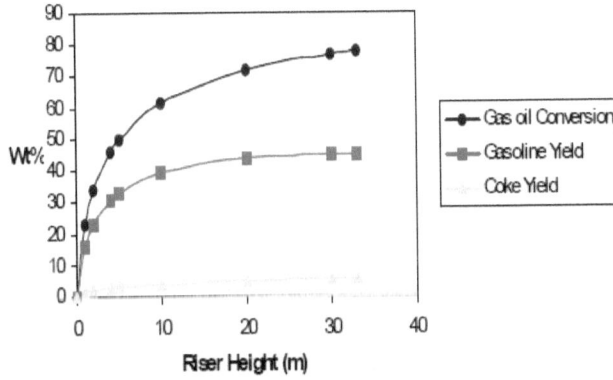

Figure 2.8: Gasoline yield and gas oil conversion in Fluid Catalytic Cracking (FCC) (Ahari *et al.*, 2008)

Fernandes *et al.* (2003) used a six-lump scheme to simulate an industrial FCC riser. 1D plug flow with negligible mass transfer resistance was also assumed by the authors. These assumptions as noted above, oversimplified the model of the authors. The product yield and temperature profiles obtained by the authors are given in Figure 2.9 and Figure 2.10 respectively.

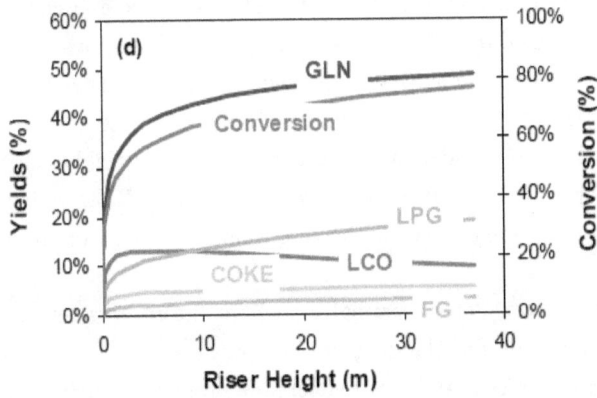

Figure 2.9: Products yield in Fluid Catalytic Cracking (FCC) (Fernandes *et al.*, 2003)

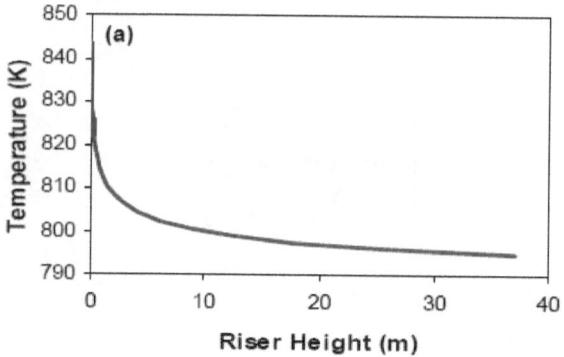

Figure 2.10: Temperature profile in Fluid Catalytic Cracking Unit (FCCU) riser (Fernandes *et al.*, 2003)

Models of higher dimensionality have also been used by other authors. Souza *et al.* (2007) used a 2D hydrodynamic, 6-lump model to simulate an industrial riser. They predicted a gasoline yield of 48%. This prediction is quite on the high side and could be attributed to the authors' negligence of mass transfer resistance in their model. Gupta (2006) and Lopes *et al.* (2012) used 3D models in their investigations. A mechanistic approach involving 50 lumps (pseudo-species) was used by Gupta (2006) to model an industrial FCCU. Lopes *et al.* (2012) on the other hand, used a 4-lump reaction scheme to investigate the effects of various exit configurations of the riser on the hydrodynamics of the reactor as well as the yield of gasoline. They found that the T-shape exit configuration enhanced the yield of gasoline owing to enhanced solid (catalyst) reflux. However, 3D models are very complex and unwieldy. They also have high costs of computation. In all the models aforementioned, the authors did not simulate the catalyst coke content thereby leaving room for more work to be done in that respect.

Pareek *et al.* (2003) developed a non-isothermal model for the riser which was incorporated in a process simulator for obtaining the temperature and conversion profiles within the riser reactor (CATCRACK). A temperature drop of about 30-40^{0}C was predicted by the authors.

A 2D model has been chosen for this work in order to circumvent the over-simplification and complexities posed by the 1D and 3D models respectively. The authors aforementioned also did not consider mass transfer resistance in their model. Hence, this research work will indubitably contribute to the existing works on riser simulation by filling the gap in the works of previous researchers.

2.6.4 Order of cracking reactions

The most commonly used order for cracking reactions in literature are first and second order rate expressions. Weekman (1968) and Weekman and Nace (1976) represented the kinetics of gas oil cracking by a second order rate expression. First order kinetics is always used for gasoline. Pitault *et al.* (1994) used first order cracking kinetics for all the lumps.

2.6.5 Riser hydrodynamics

The riser of FCCU is divided into two zones each having different functions, viz:
- Feed-injection zone
- Middle section and Upper section

Feed-injection zone:

In the feed-injection zone, the hydrocarbons feed sprayed in the form of droplets through the feed nozzles contact hot regenerated catalyst. This intimate contact rapidly vaporizes the feed and the increased amount of vapour raises the velocity and lowers the density of the flowing system. The feed-injection zone consists of three phases: catalyst (solid), hydrocarbon vapour (gas) and hydrocarbon droplets (liquid). Also in this zone, the velocity, temperature and concentration gradients are high. Only hydrocarbon vapours and solid catalysts are present in the middle and upper section of the riser. This is because all the feed droplets vaporize after traveling 2-4m up from the feed inlet (Gupta, 2006).

In a typical injector configuration, catalysts flowing at high density come in contact with feed droplets at the inlet of the reactor. About 5% steam serving as atomizing media is injected into this zone (Gupta, 2006). Merry, 1971; Chen and Weinstein (1993) showed that the feed jets introduced into a fluidized-bed reactor have a considerable effect on the hydrodynamics of the riser. The jet can form a void, bubble trains and a surrounding compaction zone (Gupta, 2006). Efforts have been made in feed-injection design to control the flow of hydrocarbons at plug flow conditions for minimizing the temperature gradients in the inlet zone which causes undesirable cracking reactions.

Middle and Upper Section:

This part of the riser has solid phase (catalyst and coke) and vapour phase (steam, hydrocarbon feed and product vapour). The catalyst particles undergo back-mixing as they rise with vapour stream. The back-mixing occurs due to slip between the solid and vapour phases. At the riser wall the velocity of the solid and vapour stream is nearly zero with the effect of back-mixing still observed. In the rest of the riser cross-section, the bulk velocity is almost the same. A core annulus type of flow pattern in circulating fluid beds has been shown to exist in several experimental studies (Gupta, 2006). Figure 2.11 gives the axial velocity profiles of the gas and the catalyst.

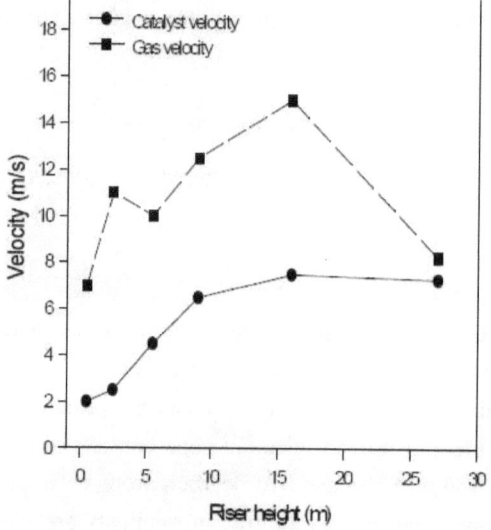

Figure 2.11: Axial velocity profiles of the gas and solid phases in the riser (Gupta, 2006)

2.6.6 Catalyst deactivation

Many types of catalytic materials are employed in the cracking of crude oil fractions. FCC catalysts consist of finely divided ($\sim 1 - 5\mu m$) lanthanide substituted X - or Y - type zeolite immeshed in amorphous silica alumina particles ($\sim 50\mu m$). The silicates may be either activated (acid treated natural clays of the bentonite type) or synthesized silica-alumina or silica-magnesia preparations. Natural and synthetic catalysts can both be used as pellets or beads and also in the form of powder. In both cases, replacements are necessary because of attrition and gradual loss of efficiency (Gerber *et al.*, 1999).

The following reasons account for the loss of activity of catalysts used for cracking:
- Structural changes due to sintering.

- Poisoning due to presence of metals (nickel and vanadium) and non-metal (sulphur, nitrogen, oxygen) in the FCC feed.
- Deposition of coke on the active sites.

Deactivation due to structural changes and metal deposition is slow and irreversible. Activity loss as a result of coke deposition is very fast but reversible. Fresh catalyst is added to the unit continually to maintain the desired activity. The circulating catalyst in the FCCU is called equilibrium catalyst (E-cat). Quantities of E-cat are periodically withdrawn from the unit (Gupta, 2006). Various models for time-dependent catalyst decay have been proposed for different lengths of contact time.

2.7 Coking of FCC Catalysts

Coke is formed where condensation of hydrocarbon vapour occurs. Higher boiling components in the feedstock readily condense and form coke nucleation sites on even slightly cooler surfaces (Speight, 2007). The high boiling feedstock constituents do not vaporize at the mixing zone of the riser. Low residence time cracking also contributes to coke deposits as there is less time for heat to transfer to feed droplets and vaporize them (Speight, 2007).

Generally, coking of fluid catalytic cracking catalysts is caused by the following factors:

i) Strong chemisorptions of molecules of desired products within the micropores of the catalyst due to high polarity and bulkiness of the molecules. The long contact time of these molecules with the inner acid sites leads to formation of coke.

ii) If blends of the feedstocks are used, excessive coking will occur if the constituents of the blends are not compatible under reactor conditions.

iii) Coke deposit also arises from the thermal decomposition of high molecular weight polar species in the feedstock (Speight, 2007).

Under normal FCC conditions, coke is the most important factor affecting catalyst activity ("Lumping and Modeling", 1996). As catalytic reactions proceed, coke covers the active sites thereby causing catalyst activity decay. In view of this, efforts have been made to model catalyst deactivation. These models are briefly discussed in the following section.

2.8 Coking Models

Froment and Bischoff (1979) proposed a mechanistic based model considering catalyst decay rate as a function of the fraction of active sites remaining and the concentration of the reactants. Corella et al. (1985) investigated the catalyst decay for a wide range of contact times (2 to 200s) and showed that the order of deactivation kinetics decreases with the contact time, taking values 3, 2, and 1 successfully.

Two major approaches have been used by authors to model catalyst activity decay. The first approach is based on catalyst coke content. The second approach relates the deactivation to catalyst time-on-stream (TOS).

2.8.1 Time-on-stream decay model

This model represents a first approach to catalyst deactivation. The approach assumes that the coking rate is independent of reactant composition, extent of conversion and hydrocarbon space velocity ("Lumping and Modeling", 1996).

Weekman (1968) used the following equations (based on TOS approach) to describe catalyst deactivation:
1. Exponential decay law:
$$\phi = exp(-\alpha t) \qquad 2.8$$
2. Power decay law:
$$\phi = t^{-n} \qquad 2.9$$
t is the catalyst time-on-stream, α and n are rate constants of the catalyst decay function.

Szepe and Levenspiel (1971) showed that the rate of catalyst activity decay can be expressed as a function of the fraction of active sites remaining as follows:
$$-\frac{d\phi}{dt} = k_d \phi^n \qquad 2.10$$
$$-\frac{d\phi}{dt} = k_{d1} C_{go}^2 \phi \qquad 2.11$$

ϕ is the fraction of remaining active sites, t is the catalyst time-on-stream, k_d and k_{d1} are catalyst decay constants, n is the order of catalyst activity decay and C_{go} is the gas oil feedstock concentration.

According to Gupta (2006), although the time-on-stream theory is widely accepted, there is no specific function which can be used for the deactivation. Various researchers employed different empirical equations to fit their experimental data. Nonetheless, the power function and the exponential function were observed to fit experimental data quite well. Kraemer et al. (1991) used the data from two different experimental reactors and showed that exponential decay function and power law function could equally represent the data. They also concluded that the simple first order decay function is effective for describing catalyst activity decay for short times (less than 20 seconds). Hence, the value of the order of deactivation used in this work is one (1) since the contact time considered is 2s.

The exponential form of the deactivation function ($\phi = exp(-\alpha t)$) was used by Kraemer and Lasa (1988) and Kraemer et al. (1990) to model oil cracking in a Chemical Reactor Engineering Centre (CREC) Riser Simulator. A modified exponential function was used to represent activity decay for the cracking of 2-methylpentane on USY zeolite. A value of 0.4 for "n" parameter was found to be adequate ("Lumping and Modeling", 1996).

2.8.2 Decay model based on coke content

Forissier and Bernard (1991) suggested an empirical relation that took into account the site coverage, pore plugging and diffusion limitations caused by pore plugging. They proposed the deactivation function given in Equation 2.12.

$$\phi = (B + 1)/(B + \exp(AC_c)) \qquad 2.12$$

C_c is the catalyst coke content expressed in wt%, A and B are two constants that are functions of feedstock composition.

Forissier et al. (1991) established a distinction between catalyst deactivation (ϕ_1) and the coke formation (ϕ_2) by the expression in Equation 2.13.

$$\phi_2 = \phi_1(C_{mc} - C_c)/C_{mc} \qquad 2.13$$

C_{mc} is the maximum coke fraction observed on the catalyst.

ϕ_2 suggests that the coke concentration, C_c can reach a limiting level C_{mc} at the early stages of the reaction. Then coke formation is halted and a residual activity is available for other cracking reactions.

Corma and Martinez-Triguero (1994) studied paraffins catalytic cracking. The authors suggested that the use of a TOS decay model is not consistent with a kinetic model in which the product olefins are strongly adsorbed on the catalyst. The authors therefore proposed the decay model given in Equation 2.14.

$$\frac{d\phi}{dt} = -k_{md}\phi^m X \qquad 2.14$$

k_{md} and m are the decay model parameters and X is the conversion.

The exponential decay model was used in this book because the catalyst activity is observed in reality to drop abruptly to an asymptotic value. The performance of a coked (0.56wt% coke on catalyst) and a fully regenerated FCC catalyst were determined by Den Hollander *et al.* (2001). The authors showed that the activity of coked catalyst is not further affected by additional coke deposition from the kinetic analysis they performed using a five-lump reaction scheme. Hence, FCC catalyst deactivation can be modeled satisfactorily by the exponential decay law.

2.9 Kaduna Refinery and Petrochemicals Company (KRPC) Ltd

Kaduna Refinery and Petrochemicals Company (KRPC) Ltd was built in 1980. The Riser Catalytic Cracking Unit (RCCU) of the plant was designed to operate with any of the three feeds given in Table 2.5 at the stated rates and conversion levels.

Table 2.5: Kaduna Refinery and Petrochemicals Company (KRPC) feedstock (Kellogg, 1980)

Feed	Crude source	Rate (m^3/day)	Conversion %
I	VQCC, Escravos, Kuwait	3342.9	80.2
II	VQCC, Escravos, Light Arabian	3347.9	80.2
III	VQCC, Escravos, Lagomar	3193.3	80.6

KRPC uses the Orthoflow "F" converter riser cracking design with a "folded" (or inverted "U" type) riser. The FCCU of KRPC consists of a Catalytic converter, CO boiler, Fractionator and Vapour Recovery Sections.

2.9.1 Catalytic cracking section

The feed to the section is Vacuum Gas Oil (VGO). The incoming feed to the section is heated initially by heat exchangers and finally by a fired heater to the riser design inlet temperature; 605K-621K (332-348^{0}C). In the riser, hot regenerated catalyst from the second stage of the regenerator (and at a temperature of at least 898K (625^{0}C)) vaporizes the feed and raises it to the reaction temperature of 796K (523^{0}C). The riser uses zeolite catalyst of average size 60 microns. The density of the equilibrium catalyst is 760kg/m^3. Cracked hydrocarbon vapours, steam and inert gas flow from the reactor to the base of the main fractionator (Kellogg, 1980).

2.9.2 Cracking riser operating temperature

In KRPC's FCCU, the hot catalyst vaporizes the feed and raises it to a reaction temperature of 796K (523^{0}C). The riser outlet temperature is controlled by the amount of catalyst admitted by the plug valve.

The reaction temperature of the riser is controlled with a temperature recording controller, which positions the regenerated catalyst plug valve to regulate the quantity of hot catalyst entering the cracking riser thereby regulating the heat transfer to the oil.

The manner in which the reaction temperature is changed in the riser brings about a change in the catalyst-to-oil ratio. Consequently, the conversion of oil feed is doubly affected by the change in reaction temperature and by the change in catalyst-to-oil ratio as the result of the change in reaction temperature. Hence, in KRPC FCC operation, it is recommended that changes in reaction temperature be made in steps of 2^0C during routine operating change in view of the dependency of catalyst-to-oil ratio on the reaction temperature (Kellogg, 1980).

The variables that affect product distribution in the riser are:
1. Catalyst activity
2. Reaction/riser outlet temperature
3. Duration of contact of oil with catalyst
4. Catalyst-to-oil ratio

The four variables above basically control the extent of conversion. In plant operation, these variables are controlled to avoid three things, viz:
1. Excessive production of gas which has lower value and is expensive to compress.
2. Excessive coke production which is also of little value and is expensive to burn off the catalyst.
3. Excessive use of fresh catalyst; the fresh catalyst is expensive.

The second objective above underscores the need for the modeling and simulation of coking of FCC catalyst. The simulation of coking in the riser reactor will help to establish the process parameters that will guarantee plant operation at reduced coking rate. Plant operation at reduced coking rate will reduce the cost of catalyst regeneration which in turn increases plant profitability. This investigation, unlike the work of previous researchers, has advanced research into coking of FCC catalyst by predicting catalyst coke content as a function of reaction temperature. Table 2.6 shows the feedstock and product design data for Kaduna Refinery and Petrochemicals Company (KRPC) Riser Catalytic Cracking Unit (RCCU).

Table 2.6: Feedstock and product design data for Kaduna Refinery and Petrochemicals Company (KRPC) Riser Catalytic Cracking Unit (RCCU) (Kellogg, 1980)

Feed stock	II	III
Fresh feed rate (kg/hr)	127,917	122,206
Products		
Gasoline, wt%	50	51

Light cycle oil, wt%	15	15
Gas, wt%	24	18

2.10 Riser Model Development

The FCCU reactor is typically a pseudo plug flow reactor. Souza *et al.* (2007) used a 2D model to simulate the FCC riser while Gupta (2006) and Lopes *et al.* (2012) used 3D models in their work. However, the assumptions usually made by previous researchers in developing their models are (Ahari *et al.*, 2008):

1. One dimensional transport and plug flow.
2. The viscosity of the feed and the heat capacity of all the components are constant.
3. Pressure changes throughout the riser length is due to the static head of catalyst in the riser.
4. In each section of the riser, the catalyst and gas are at thermal equilibrium.
5. The coke has the same specific heat as the catalyst.
6. The riser dynamic is fast enough to justify a quasi-steady state model.
7. At the riser inlet, hydrocarbon feed comes into contact with the hot catalyst coming from the regenerator and instantly vaporizes (taking away latent heat and sensible heat from the hot catalyst (Gupta, 2006).
8. There is no loss of heat from the riser and the temperature of the reaction mixture (hydrocarbon vapours and catalyst) falls only because of the endothermicity of the cracking reactions (Gupta, 2006).
9. Ideal gas law is assumed to hold.
10. Mass transfer resistance is negligible.

The assumptions in contention in this book are the first and the last assumptions (1D plug flow and negligible mass transfer resistance).

2.11 COMSOL Multiphysics and MATLAB

COMSOL Multiphysics formally known as FEMLAB, is a Finite Element Analysis (FEA) and solver software and package for diverse Physics and Engineering applications. The software was started by graduate students to Germund Dahlquist based on codes developed for a grand course at the Royal Institute of Technology in Stokholm, Sweden (National Agency for Science and Engineering Infrastructure [NASENI], 2010). COMSOL has an interface to MATLAB apart from its own interface, COMSOL Script (Simonsen, 2008).

2.11.1 Unique features of COMSOL Multiphysics

COMSOL Multiphysics can be used to solve coupled systems of partial differential equations (PDEs). The software also comes handy with a mathematical interface that solves systems of differential algebraic equations (DAEs).

The power of COMSOL Multiphysics as a modeling software is evident in the number of application-specific modules that the software has. These modules are:
a) AC/DC Module
b) Acoustic Module
c) CAD import Module
d) Chemical Engineering Module (which incorporates Reaction Engineering Laboratory application mode).
e) Earth Science Module
f) Heat Transfer Module
g) Material Library
h) Microelectromechanical Systems Module (MEMS)
i) Structural Mechanics Module

2.11.2 Reaction engineering laboratory (REL)

Reaction Engineering Laboratory (REL) is an application mode in COMSOL Multiphysics that is used to model and simulate reactions and reactors with amazingly impressive speeds. The beauty and power of Reaction Engineering Laboratory mode of COMSOL Multiphysics is seen in its various uses which include:
a) Modeling of chemical reactions (species concentration transients for both single and multiple reactions).
b) Determination of reaction parameters such as rate constants, reaction order and Arrhenius parameters from experimental data.
c) Reactor modeling (time and space modeling of reactor geometry for batch reactor, Continuous Stirred Tank Reactor (CSTR), Plug Flow Reactor (PFR), and fixed bed reactor).

2.11.3 Unique features of MATLAB

MATLAB (Matrix Laboratory) is also another powerful mathematical tool for solving mathematical problems/models. Examples of models that can be solved in MATLAB are:
a) Linear and non – linear systems of equations.

b) Systems of Ordinary Differential Equations (ODEs) using different ODE solvers such as ode15, ode23s etc depending on the stiffness of the system.

c) MATLAB also allows users to write their own codes by using the software's function and M-files.

2.12 Derivation of Model Rate Equation

The riser model here presented is be based on the following assumptions:

a) Pseudo homogenous two-dimensional transport with axial and radial gradients. In reality the riser is a 3D reactor. Simplifying the geometry to 1D is tantamount to predicting products yield just along the axis of the reactor. However turbulent the flow in the riser may be, a 1D model cannot adequately represent the entire geometry of the reactor because it does not account for wall effects. 3D riser models on the other hand are unwieldy and costly in terms of computational time and memory. A 2D model however, takes into account the wall effects of the reactor and is neither as over-simplified as the 1D models nor as unwieldy (complicated) as the 3D models.

b) In each section of the riser, the catalyst and gas are at thermal equilibrium.

c) The riser dynamic is fast enough to justify a quasi-steady state model.

d) At the riser inlet, hydrocarbon feed comes into contact with the hot catalyst coming from the regenerator and instantly vaporizes (Gupta, 2006). (The feed is preheated to its boiling point before it is injected into the riser through nozzles. The feed aerosols and the hot catalyst introduced at the bottom of the riser enhance instantaneous vaporization of feed.)

e) There is no loss of heat from the riser and the temperature of the reaction mixture (hydrocarbon vapours and catalyst) falls only because of the endothermicity of the cracking reactions (Gupta, 2006). (The inner wall of the riser is known to be lined with refractory material.)

FCC catalysts are porous in nature. The porous nature of the catalyst particle gives rise to significant gradients of concentration and temperature across the particle.

The concentration and temperature gradients in zeolite is as illustrated in Figure 2.12 for the general first order catalytic reaction given in Equation 2.15.

$$A_{(g)} \rightarrow products \qquad 2.15$$

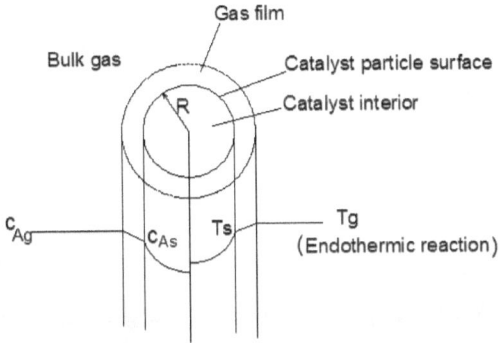

Figure 2.12: Concentration and temperature gradients in a porous catalyst (Missen *et al.*, 1999)

c_{As} = concentration of A on catalyst surface
c_{Ag} = concentration in bulk gas
T_s = temperature at surface of catalyst
T_g = bulk gas temperature

2.12.1 Effective diffusivity

Diffusion (spontaneous migration of specie in space relative to other species, as a result of a variation in its chemical potential, in the direction of decreasing potential) can occur inside a porous catalyst particle in three different modes, viz: molecular, Knudsen and surface diffusions.

Molecular diffusion occurs as a result of molecular collisions in pores (for particles with relatively large pores and at high pressure). Knudsen diffusion is dominant in small pores at low pressure and it results from collision of molecules with the walls of the catalyst pores. Surface diffusion mechanism is as a result of the migration of adsorbed species along the pore surface due to surface concentration gradient. Molecular and Knudsen diffusion mechanisms were considered in deriving the reaction rate equation in this work. (Surface diffusion mechanism has been neglected because FCC reactions occur instantaneously on the catalyst surface.)

Fick's law:

The steady state 1D diffusion of a species A with molar flux, N_A in molm^{-2}(cross-sectional area of diffusion medium)s^{-1} through a particle is given by Fick's law in Equation 2.16.

$$N_A = -D_e \frac{\partial c_A}{\partial z} \qquad 2.16$$

D_e is the effective diffusivity. The effective diffusivity can be estimated in terms of molecular diffusivity D_{AB}, knudsen diffusivity, D_k, particle voidage ε_p and tortuosity τ_p. D_{AB} can be estimated from Equation 2.17 (Geankoplis, 2011).

$$D_{AB} = \frac{1.8583 \times 10^{-7} T^{\frac{3}{2}} \left[(M_A + M_B)/M_A M_B \right]^{\frac{1}{2}}}{P \Omega_D d_{AB}^2} \qquad 2.17$$

D_{AB} in m^2/s, T in K, M_A and M_B are molecular masses in $kgmol^{-1}$, P in atm, d_{AB} is the collision diameter which is evaluated as the average of the molecular diameters of the colliding species, $(d_A + d_B)/2$ in nm while Ω_D is the collision integral. Knudsen diffusivity on the other hand is given by the correlation in Equation 2.18 (Missen et al., 1999).

$$D_k = 97 r_e \left(T/M \right)^{\frac{1}{2}} \qquad 2.18$$

D_k in m^2/s, r_e is the average pore radius in m. The equation holds for all geometries except straight, cylindrical pores.

Overall diffusivity, D^*:

The overall diffusivity and the effective diffusivity can be evaluated from the correlations given in Equation 2.19 and Equation 2.20 (Missen et al., 1999).

$$\frac{1}{D^*} = \frac{1}{D_{AB}} + \frac{1}{D_k} \qquad 2.19$$

$$D_e = \frac{D^* \varepsilon_p}{\tau_p} \qquad 2.20$$

D_e in $m^3 (void\ space) m^{-1} (particle) s^{-1}$, tortuosity, τ_p ranges between 3 and 7 for most catalysts.

Thiele modulus, φ:

Thiele modulus, φ (ratio of intrinsic reaction rate to diffusion rate through De) is given by Equation 2.21 for spherical catalyst particles.

$$\varphi = R \left(\frac{k_A}{D_e} \right)^{\frac{1}{2}} \qquad 2.21$$

Equation 2.21 provides a yardstick for determining the rate determining step in solid catalysis; whether it is chemical reaction via k_A or diffusion via D_e.

The particle effectiveness factor, η:

Generally, for a steady state diffusion of species A through catalyst pores and for the first order reaction given by Equation 2.15, the particle effectiveness factor, η is given by Equation 2.22a:

$$\eta = \frac{\text{rate with diffusion resistance}}{\text{rate with no diffusion resistance}} = \frac{(-r_A)_{observed}}{(-r_A)_{intrinsic}} \quad 2.22a$$

$$(-r_A)_{intrinsic} = k_A c_{As} \quad 2.22b$$

For a first order reaction $(-r_A)$ is in kmolAm^{-3}(particle)s^{-1} or kmolAkg^{-1}(particle)s^{-1}.

For spherical particles (Missen *et al.*, 1999):

$$\eta = \frac{3}{\varphi}\left(\frac{1}{\tanh\varphi} - \frac{1}{\varphi}\right) \quad 2.23$$

The particle effectiveness factor is a direct measure of the extent to which diffusion resistance reduces the rate of chemical reactions in solid catalysis and it is a function of Thiele modulus.

Overall effectiveness factor, η_0:

The particle effectiveness factor only considers the concentration gradient within the particle but neglects the gradient from bulk fluid to the exterior surface of the particle. The overall effectiveness factor, η_0 incorporates the bulk gas condition, c_{Ag}.

$$\eta_0 = \frac{r_A(observed)}{r_A(c_{Ag})} \quad 2.24$$

From Equations 2.22 and Equation 2.24,

$$(-r_A) = \eta k_A c_{As} \quad 2.25$$
$$(-r_A) = \eta_0 k_A c_{Ag} \quad 2.26$$

Also at steady state:

$$(-r_A) = k_g(c_{Ag} - c_{As}) \quad 2.27$$

The expression for particle overall effectiveness factor, Equation 2.28 is arrived at by eliminating c_{As} from Equation 2.26 and Equation 2.27.

$$\eta_0 = \frac{1}{\left(\frac{k_A}{k_g}\right) + \left(\frac{1}{\eta}\right)} \quad 2.28$$

Substituting Equation 2.28 in Equation 2.26 gives the model rate equation.

$$(-r_A) = \frac{a c_{Ag}}{\left(\frac{1}{k_g}\right) + \left(\frac{1}{\eta k_A}\right)} \quad 2.29$$

The basic parameters to be determined in Equations 2.21 and 2.29 are D_e and k_g.

Estimation of mass transfer coefficient, k_g:

The mass transfer coefficient, k_g in Equation 2.29 can be estimated from Sherwood number, N_{sh} for gases which is given by Equation 2.30 (Geankoplis, 2011).

$$N_{sh} = 2 + 0.552 N_{Re}^{0.53} N_{sc}^{1/3} \qquad 2.30$$

$$N_{sc} = 0.6 - 2.7 \text{ and } N_{Re} = 1 - 48\,000$$

$$N_{sh} = \frac{k_g D_p}{D_{AB}} \qquad 2.31$$

$$N_{sc} = \text{Schmidt number} = \frac{\mu}{\rho D_{AB}} \qquad 2.32$$

$$N_{Re} = \text{Particle Reynolds number} = \frac{D_p v \rho}{\mu} \qquad 2.33$$

2.12.2 Modeling FCC reactions using the five-lump model

In the five-lump model given in Figure 2.7, the eight reactions of the model are taken to follow first order kinetics as follows (Den Hollander *et al.*, 2003):

$$r_j = k_j c_i \quad j = 1,\ldots,8 \qquad 2.34$$

The modeling is carried out in two stages in COMSOL Multiphysics. Plug-flow reactor is assumed at a temperature of 791K. The coking reactions (reactions 7 and 8) were neglected in the first stage whereas in the second stage, they were taken into consideration. Equation 2.35 gives the mass balance for the reactions.

$$\frac{dF_i}{dV} = \sum_j v_{ij} r_j = r_i \qquad 2.35$$

F_i is the species mass flow rate, V is the reactor volume, v_{ij} is the stoichiometric coefficient and r_i the rate of production/consumption of species i.

The species concentration gradient as a function of residence time (t) is derived from Equation 2.35 and expressed in Equation 2.36 below.

$$\frac{dF_i}{dV} = \frac{d(vc_i)}{dV} = \frac{dc_i}{dt} = r_i \qquad 2.36$$

The lumped species mass balance for the first simulation (the model without coking) is given by Equations 2.37.

$$\frac{dc_{vgo}}{dt} = -r_1 - r_2 - r_3 \qquad 2.37a$$

$$\frac{dc_{lco}}{dt} = r_1 - r_4 - r_5 \qquad 2.37b$$

$$\frac{dc_{gasoline}}{dt} = r_2 + r_5 - r_6 \qquad 2.37c$$

$$\frac{dc_{gas}}{dt} = r_3 + r_4 + r_6 \qquad 2.37d$$

When coking reactions are considered, the overall activity of the system is dominated by two different time scales. This is because coke formation takes place on the millisecond scale while the formation of the other products occurs in seconds. This challenge can be overcome in COMSOL Multiphysics by using two different activity functions, a and b for the non-coking and the coking reactions respectively. Equation 2.38 was used when coking was considered in the second simulation. A coke concentration-based decay model, 'a' given by Equation 2.39 can be used for the non-coking reactions. The function, b on the other hand will be based on catalyst time-on-stream decay model (Den Hollander et al., 2003). Hence for the second simulation, with mass transfer resistance taken into consideration, the reaction rates become:

$$r_j = \frac{ac_i}{\left(\frac{1}{k_g} + \left(\frac{1}{\eta k_j}\right)\right)} \qquad j = 1, \dots, 6 \qquad 2.38$$

$$a = \exp(-k_d c_{coke}) \qquad 2.39$$

$$r_j = \frac{bc_i}{\left(\frac{1}{k_g} + \left(\frac{1}{\eta k_j}\right)\right)} \qquad j = 7, 8 \qquad 2.40$$

$$b = \exp(-\alpha' t) \qquad 2.41$$

$k_d = 8.2, \alpha' = 10s^{-1}$ (Den Hollander et al., 2003)
k_g = mass transfer coefficient of reactant in m/s, η = particle effectiveness factor.

The reactor lumped species balances with the coking reactions included were given by Equations 2.42.

$$\frac{dc_{vgo}}{dt} = -r_1 - r_2 - r_3 - r_7 \qquad 2.42a$$

$$\frac{dc_{lco}}{dt} = r_1 - r_4 - r_5 - r_8 \qquad 2.42b$$

$$\frac{dc_{gasoline}}{dt} = r_2 + r_5 - r_6 \qquad 2.42c$$

$$\frac{dc_{gas}}{dt} = r_3 + r_4 + r_6 \qquad 2.42d$$

$$\frac{dc_{coke}}{dt} = r_7 + r_8 \qquad 2.42e$$

2.12.3 Development of two-dimensional pseudo homogeneous riser model

The continuity and energy equations of the reactor can be developed using the control volume (shell element) shown in Figures 2.13.

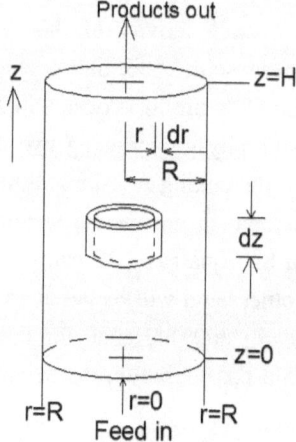

Figure 2.13(a): 2D riser

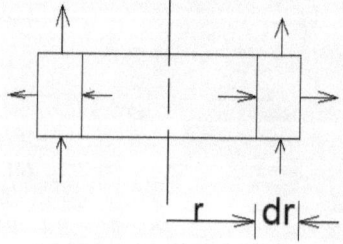

Figure 2.13(b): Control volume for riser model

Continuity equation:

Material balance for species i around the control volume is given by Equation 2.43:

$(Input\ of\ i\ by\ bulk\ flow + axial\ dispersion + radial\ dispersion)$
$= (Output\ rate\ of\ i\ by\ bulk\ flow + axial\ dispersion + radial\ dispersion + chemical\ reaction\ )$ 2.43

$$uc_i 2\pi r dr - D_{zi} 2\pi r dr \frac{\partial c_i}{\partial z} - D_{ri} 2\pi r dz \frac{\partial c_i}{\partial r}$$
$$= 2\pi r dr \left[uc_i + \frac{\partial(uc_i)}{\partial z} dz\right] - D_{zi} 2\pi r dr \left[\frac{\partial c_i}{\partial z} + \frac{\partial}{\partial z}\left(\frac{\partial c_i}{\partial z}\right) dz\right]$$
$$- D_{ri} 2\pi dz \left[r\frac{\partial c_i}{\partial z} + \frac{\partial}{\partial z}\left(r\frac{\partial c_i}{\partial z}\right) dr\right] + \rho_B(-r_i) 2\pi r dr dz$$

$$D_{zi}\frac{\partial^2 c_i}{\partial z^2} + D_{ri}\left(\frac{\partial^2 c_i}{\partial r^2} + \frac{1}{r}\frac{\partial c_i}{\partial r}\right) - \frac{\partial(uc_i)}{\partial z} - \rho_B(-r_i) = 0 \qquad 2.44$$

$$u = \frac{q}{A_c}, m^3(fluid)s^{-1}m^2(vessel) \qquad 2.45$$

q is the volumetric flow rate of the gas through interparticle bed voidage, $m^3(fluid)s^{-1}$, D_z and D_r are effective diffusivities in $m^3(fluid)m^{-1}(vessel)s^{-1}$, $(-r_i)$ is in $kg\ species\ kg^{-1}(catalyst)\ s^{-1}$. The boundary conditions associated with Equation 2.44 are:

@ $z = 0, 0 < r < R$ (inlet): $c_{vgo} = c_0, c_{lco} = c_{gasoline} = c_{gas} = c_{coke} = 0$ \qquad 2.46

@ $z = H, 0 < r < R$ (outlet): $\frac{\partial c}{\partial z} = 0$ \qquad 2.47

@ $r = 0, 0 < z < H$: $\frac{\partial c}{\partial r} = 0$ \qquad 2.48

@ $r = R, 0 < z < H$: $\frac{\partial c}{\partial r} = 0$ \qquad 2.49

z is the axial direction, r is the radial direction, R = radius of riser (in m), H = riser height (in m), c_0 = initial concentration of VGO in mass fraction.

Hydrodynamic model of the riser reactor:

The numerical value of the catalyst slip factor (the ratio of the gas interstitial velocity to the average particle velocity) can be predicted from Equation 2.50 (Ahari et al., 2008):

$$\psi = \frac{u_0}{\varepsilon v_p} = 1 + \frac{5.6}{Fr} + 0.47 Fr_t^{0.47} \qquad 2.50$$

Fr = Froude number given by Equation 2.51 and Fr_t = Froude number at terminal velocity.

$$F_r = \frac{u_0}{(gD)^{0.5}} \qquad 2.51$$

g = acceleration due to gravity (m²/s).

The average particle velocity in the riser, v_p is given by Equation 2.52.

$$v_p = \frac{G_s}{\rho_s(1-\varepsilon)} \qquad 2.52$$

G_s is the catalyst mass flux.

Equation 2.53 for the average voidage in terms of the solid mass flux, superficial gas velocity, riser diameter and catalyst physical properties was derived from Equation 2.50 and Equation 2.52.

$$\varepsilon = 1 - \frac{G_s\psi}{u_0\rho_s + G_s\psi} \qquad 2.53$$

In KRPC Riser Catalytic Cracking unit (RCCU), a catalyst pick-up velocity of above 4.27m/s is required in the riser to prevent "slugging" of catalysts (Kellogg, 1980).

Energy balance:

The energy balance for species i around the control volume (Figure 2.13(b)) is given by Equation 2.54:

(Enthalpy input rate by bulk flow + axial conduction + radial conduction + input by chemical reaction)
= (Enthalpy output rate by bulk flow + axial conduction + radial conduction)
$\qquad 2.54$

$$G_s 2\pi r dr c_p (T - T_{ref}) - k_z 2\pi r dr \frac{\partial T}{\partial z} - k_r 2\pi r dz \frac{\partial T}{\partial r} + \rho_B(-r_i)(-\Delta H_{Ri}) 2\pi r dr dz$$
$$= G_s 2\pi r dr c_p \left[(T - T_{ref}) + \frac{\partial T}{\partial z} dz\right] - k_z 2\pi r dr \left[\frac{\partial T}{\partial z} + \frac{\partial}{\partial z}\left(\frac{\partial T}{\partial z}\right) dz\right]$$
$$- k_r 2\pi dz \left[r\frac{\partial T}{\partial r} + \frac{\partial}{\partial r}\left(r\frac{\partial T}{\partial r}\right) dr\right]$$

$$k_z \frac{\partial^2 T}{\partial z^2} + k_r \left(\frac{\partial^2 T}{\partial r^2} + \frac{1}{r}\frac{\partial T}{\partial r}\right) - G_s c_p \frac{\partial T}{\partial z} + \rho_B \sum_{i=1}^{8}(-r_i)(-\Delta H_{Ri}) = 0 \qquad 2.55$$

k_z and k_r are the effective thermal conductivities.

The boundary conditions associated with Equation 2.55 are:

@ $z = 0, 0 < r < R \ (inlet): T = T_0$ $\qquad 2.56$

@ $z = H, 0 < r < R \ (outlet): \frac{\partial T}{\partial z} = 0$ $\qquad 2.57$

@ $r = 0, 0 < z < H: \frac{\partial T}{\partial r} = 0$ $\qquad 2.58$

@ $r = R, 0 < z < H: \frac{\partial T}{\partial r} = 0$ $\qquad 2.59$

z is the axial direction, r is the radial direction, R = radius of riser (in m), H = riser height (in m), T_0 = inlet temperature of the riser.

The coupling between the riser and the regenerator is expressed in the model by Equation 2.60 (Ahari et al., 2008):

$$F_{cat}C_{pcat}(T_0 - T_{cat}) + F_f C_{pfl}(T_{vap} - T_f) + F_f C_{pfv}(T_0 - T_{vap}) + F_f \Delta H_{vap} = 0 \quad 2.60$$

The catalyst-to-oil ratio can be calculated from Equation 2.60 as a function of reactor inlet temperature.

CHAPTER 3
MATERIALS AND METHODOLOGY
3.1 Modeling Strategy

The FCCU reactor was modeled in this work using COMSOL Multiphysics software (Version 4.0) and MATLAB (R2009a) on a Compaq HP CQ61 laptop. The modeling was carried out in two stages:
- FCC reactions modeling (done in COMSOL Multiphysics).
- Reactor/Geometry modeling (done in MATLAB).

3.1.1 FCC reaction modeling

The five-lump model FCC reactions were modeled in COMSOL 4.0 as follows:
1. COMSOL Multiphysics was launched.
2. The preferences were set by selecting Reaction Engineering application mode (time-dependent study type).
3. Reaction Engineering was selected in the model tree and the reaction temperature (791K) was entered.
4. In the Reaction Tab (in the model tree), six reactions were added (the non-coking reactions with rate constants k_1 to k_6 depicted in Figure 2.5).
5. Each of the reactions was selected and the actual irreversible reaction was entered in the Setting's page. The corresponding forward rate constants for the reactions were also entered.
6. The global parameters (constants) for the reaction modeling (α' and k_d in Equations 2.39 and 2.41) were entered.
7. The global expressions (a and b from Equation 2.39 and Equation 2.41) were then entered.
8. The Species tab was expanded and each species' molecular weights and initial concentrations were entered in the Settings page.
9. The sequence was generated from the study and the results were computed from the sequence so generated. This completed the modeling of the non-coking reactions.

To model the FCC reactions with coking taken into consideration:
1. Two more reactions were added in the model tree.
2. The two reactions that lead to coke via k_7 and k_8 (Figure 2.5) were entered in the settings page.
3. The reaction rate expression for each of the non-coking reactions was multiplied by the concentration-dependent deactivation function 'a'.

3. The rates of the coking reactions were multiplied by the time-dependent deactivation expression, b (Equation 2.41).
4. Finally, the sequence was regenerated from the study and the solution was computed again to model both the coking and non-coking reactions.

The concentrations of the species involved in the FCC reactions were modeled as a function of time in the Reaction Engineering application mode.

3.1.2 Modeling of riser

The governing equations (Equation 2.44 and Equation 2.55) were expressed in a general, normalised form as follows:

$$\alpha \left(\frac{\partial^2 \sigma}{\partial r^{*2}} + \frac{1}{r^*} \frac{\partial \sigma}{\partial r^*} \right) + \beta \frac{\partial^2 \sigma}{\partial z^{*2}} + \gamma \frac{\partial \sigma}{\partial z^*} + \lambda(-r_i) = 0 \qquad 3.1$$

$$\sigma = {c_i}/{c_0} \text{ or } {T}/{T_0} \qquad 3.2$$

$$r^* = {r}/{R} \qquad 3.3$$

$$z^* = {z}/{H} \qquad 3.4$$

The coefficients in Equation 3.1 are given by the following expressions:

$$\alpha_1 = \alpha_2 = 1 \qquad 3.5$$

$$\beta_1 = \frac{R^2 D_z}{H^2 D_r} \qquad 3.6$$

$$\beta_2 = \frac{R^2 k_z}{H^2 k_r} \qquad 3.7$$

$$\gamma_1 = \frac{-G c_p R^2}{H k_r} \qquad 3.9$$

$$\lambda_1 = \frac{R^2 \rho_B}{D_r c_0} \qquad 3.10$$

$$\lambda_2 = \frac{R^2 \rho_B}{k_r T_0} \qquad 3.11$$

Subscripts 1 and 2 in the coefficients correspond to the continuity equation and energy balance respectively.

Boundary conditions for the normalized equation (Equation 3.1):

@ $z^* = 0, 0 < r^* < 1$: $\sigma_{vgo} = \sigma_T = 1, \sigma_{lco} = \sigma_{gasoline} = \sigma_{gas} = \sigma_{coke} = 0$ 3.12

@ $z^* = 1, 0 < r^* < 1$ (outlet): $\frac{\partial \sigma}{\partial z^*} = 0$ 3.13

@ $r^* = 0, 0 < z^* < 1$: $\frac{\partial \sigma}{\partial r^*} = 0$ 3.14

$$@\ r^* = 1, 0 < z^* < 1:\ \frac{\partial \sigma}{\partial r^*} = 0 \qquad 3.15$$

3.1.3 Finite difference discretization of the governing equations

The finite difference grid shown in Figure 3.1 was used to discretize and solve the governing equations; Equation 2.44 and Equation 2.55.

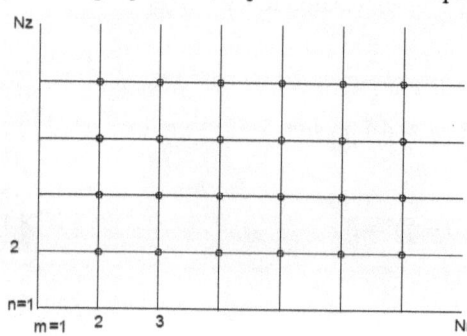

Figure 3.1: Finite difference grid

The vertical axis corresponds to the distance along the height of the riser while the horizontal axis of the grid corresponds to the radial direction.

Centered difference scheme was used to discretize the second order differentials with the aid of the graphical representation shown in Figure 3.2.

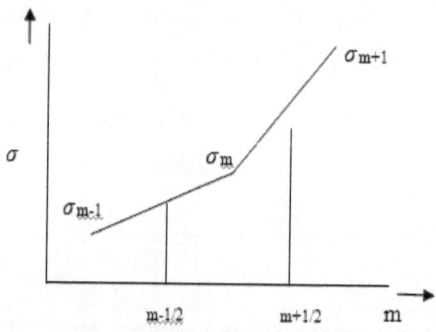

Figure 3.2: Discretization of second order differentials

The resulting discretised equation is given by Equation 3.16:

$$\left(\frac{\beta}{(\Delta z)^2}+\frac{\gamma}{\Delta z}\right)\sigma_{m,n+1} - \left(\frac{2\alpha}{(\Delta r)^2}+\frac{2\beta}{(\Delta z)^2}+\frac{\alpha}{r\Delta r}+\frac{\gamma}{\Delta z}\right)\sigma_{m,n} + \frac{\alpha}{(\Delta r)^2}\sigma_{m-1,n}$$
$$+ \left(\frac{\alpha}{(\Delta r)^2}+\frac{\alpha}{r\Delta r}\right)\sigma_{m+1,n} + \frac{\beta}{(\Delta z)^2}\sigma_{m,n-1} + \lambda(-r_i)|_{m,n} = 0 \qquad 3.16$$

Equation 3.16 was expressed in terms of k^* (actual node numbers) for ease of coding as follows:

$$\left(\frac{\beta}{(\Delta z)^2}+\frac{\gamma}{\Delta z}\right)\sigma_{k^*+1} - \left(\frac{2\alpha}{(\Delta r)^2}+\frac{2\beta}{(\Delta z)^2}+\frac{\alpha}{r\Delta r}+\frac{\gamma}{\Delta z}\right)\sigma_{k^*} + \frac{\alpha}{(\Delta r)^2}\sigma_{k^*-1}$$
$$+ \left(\frac{\alpha}{(\Delta r)^2}+\frac{\alpha}{r\Delta r}\right)\sigma_{k^*-Nr} + \frac{\beta}{(\Delta z)^2}\sigma_{k^*+Nr} + \lambda(-r_i)|_{k^*} = 0 \qquad 3.17$$

$$k^* = N_r(n-1) + m \qquad 3.18$$

Grid optimization was carried out by using 10x10, 20x20, 50x50 and 100x100 grids to solve the discretized equation. The model was finally, solved by using 20x20 grid because it converged better and gave more accurate results than the other grid sizes. The second modeling was carried out to simulate the reactor by modeling the flow rates of the species involved in the reaction as a function of reactor height.

The following steps were taken to model the reactor geometry in MATLAB:

1. MATLAB (2009a) was launched.
2. A new M - file was created for the discretized system of algebraic equations (6 equations).
3. The relevant constants and variables were declared and initialized.
4. The program to solve the riser model equations and display the results was written in the M-file.

Figure 3.3 shows the sequence of work for the simulation of the riser. The coking of the catalyst was modeled in COMSOL Multiphysics (reaction modeling) using the five-lump model of Den Hollander et al. (2003). The modeling of the FCC reactions was done to obtain the concentration profiles (evolution of species over time) of the species. Thereafter, the reactor was modeled (geometry modeling) by using the conservation laws; mass and energy balances. The governing equations that were derived for the riser were then solved in MATLAB to predict the concentrations of species along the length of the reactor.

The model predictions were then validated by comparing them with data obtained from literature as well as from Kaduna Refinery and Petrochemical Company (KRPC) Ltd. Finally, the validated model was used to simulate the coking of FCC catalyst within a temperature range of 779K to 791K.

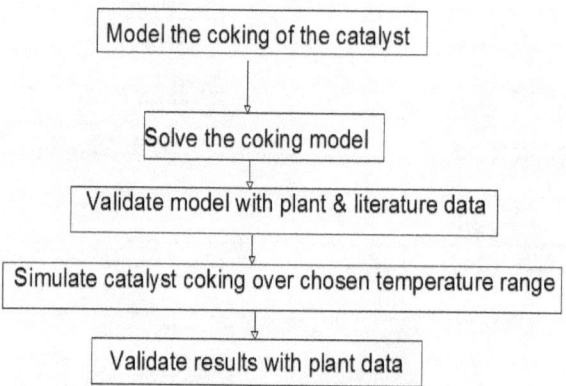

Figure 3.3: Sequence of work for the simulation of Fluid Catalytic Cracking (FCC) reactions

Figure 3.4 depicts the computational flow diagram for the riser model here presented. The two basic stages of the work are as shown in the diagram. The first stage involved modeling the FCC reactions whereas in the second stage, the FCC reactor was modeled to predict the axial and radial variation in species concentrations within the riser.

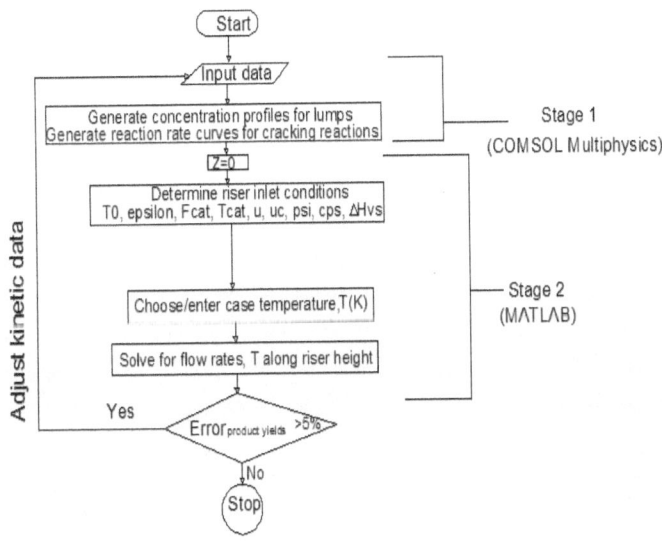

Figure 3.4: Computational flow diagram for riser model

3.2 Riser Kinetic Data

Table 3.1, Table 3.2 and Table 3.3 give the thermodynamic properties of FCC reactions and species properties as obtained from literature.

Table 3.1: Enthalpies of cracking (Ahari *et al.*, 2008)

S/N	Cracking reaction	ΔH(kJ/kg)
1	VGO to LCO	80
2	VGO to gasoline	195
3	VGO to gas	670
4	LCO to gas	-
5	LCO to gasoline	180
6	Gasoline to gas	530
7	VGO to coke	745
8	LCO to coke	600

Table 3.2: Molecular weights and heat capacities (Ahari *et al.*, 2008)

S/N	Species	Molecular weight (kg/kmol)	Cp (kJ/kg.K)
1	VGO	333.0	2.67 (liquid), 3.30 (gas)
2	LCO	300.0	3.30
3	Gasoline	106.7	3.30

4	Gas	40.0		3.30
5	Coke	14.4		1.087

Table 3.3: Gas oil properties

Property	Value	Source
Specific gravity	0.89-0.93	Gupta (2006)
Sulphur, wt%	0.30-2.00	Gupta (2006)
Nitrogen, wt%	0.07-2.00	Gupta (2006)
Conardson carbon, wt%	0.00-0.20	Gupta (2006)
Viscosity	$1.4 \times 10^{-5} N.s/m^2$	Ahari et al. (2008)
Vaporization temperature	698K	Ahari et al. (2008)
Enthalpy of vaporization	190kJ/kg	Ahari et al. (2008)

The process parameters used in the model are given in Table 3.4.

Table 3.4: Riser model parameters

Parameter	Value	Source
Reactor inlet temperature, T_0 (K)	791	KRPC
Feed inlet temperature, T_f (K)	613	KRPC
Catalyst inlet temperature, T_{cat} (K)	927	KRPC
Specific heat capacity (liquid feed), cpfl (J/kg-K)	2.67e3	Ahari et al. (2008)
Specific heat capacity (vapour feed), cpvf (J/kg-K)	3.30e3	Ahari et al. (2008)
Specific heat capacity (catalyst), cpcat (J/kg-K)	1.09e3	Ahari et al. (2008)
Feed vaporization temperature, T_{vap} (K)	698	KRPC
Enthalpy of vaporization, delHvap (J/kg)	190e3	Ahari et al. (2008)
Density (solid catalyst), ρ_s (kg/m^3)	1250	KRPC
Catalyst velocity, U_c (m/s)	4.8	Gupta (2006)
Gas superficial velocity, U (m/s)	18	KRPC
Slip factor, psi	2	KRPC
Feed flow rate, F_f (kg/s)	35.5	KRPC
Riser diameter, D_R (m)	1.146	KRPC
Riser height, H (m)	25	KRPC
Pore diameter, P_d (m)	2.00e-9	KRPC
Particle diameter, D_p (m)	60e-6	KRPC
Gas average density ρ_g (kg/m^3)	0.92	KRPC
Gas average viscosity μ_g (Pa.s^{-1})	1.40e-5	Ahari et al. (2008)
Riser pressure, P (atm)	2.94	KRPC
Particle tortuosity, τ_p	7	Missen et al. (1999)

CHAPTER 4
RESULTS AND DISCUSSION

This chapter presents and discusses the results obtained in COMSOL Multiphysics and MATLAB. The results obtained in COMSOL Multiphysics were discussed in Section 4.1, while the results from MATLAB were presented and analyzed in Section 4.2 and Section 4.3.

4.1 Reaction Model Results (COMSOL Multiphysics)

The concentration-time plots for the cracking reactions are shown in Figures 4.1 and 4.2. Figure 4.1 shows the concentration profiles of species when only mass transfer resistance was considered in the FCC reactions model. The result from literature that is given in Figure 2.6 depicts the effect of coking alone. Figure 4.2 on the other hand depicts the effects of both coking and mass transfer resistance on products yield and selectivity. Figure 4.3 graphically depicts the difference between the model predictions and the predictions from literature by Den Hollander *et al.* (2003). Figure 4.4 depicts the decline in catalyst activity for the reactions (profile 'a') as well as the concentration profile of coke formation within the simulation time.

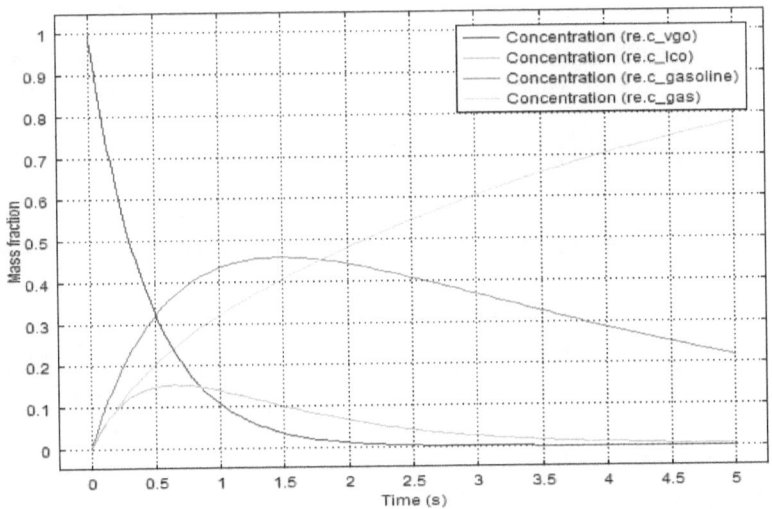

Figure 4.1: Effect of mass transfer resistance on Fluid Catalytic Cracking (FCC) reactions (no coking)

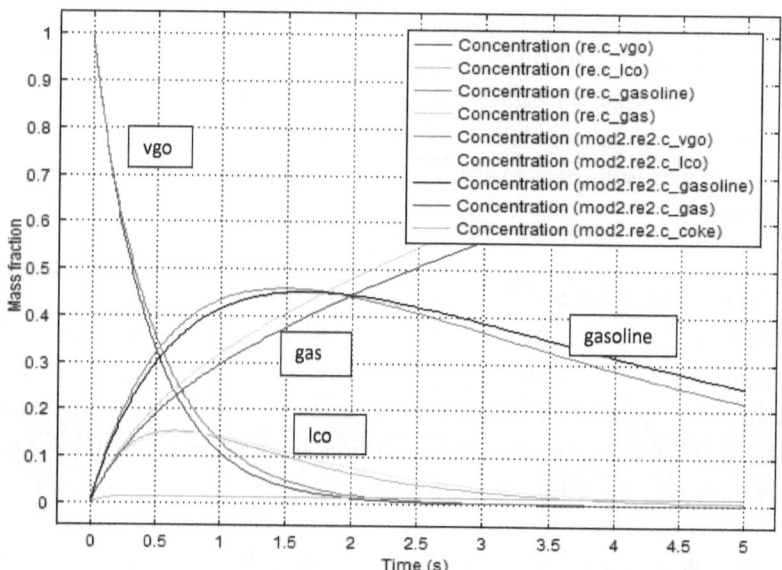

Figure 4.2: Effects of mass transfer resistance and catalyst coking on selectivity and yield of Fluid Catalytic Cracking (FCC) (vgo: Vacuum gas oil. lco: Light cycle oil. re.c_i: COMSOL notation for species concentrations obtained in reaction engineering (re). mod2.re2.c_i: species concentrations for the second model in reaction engineering)

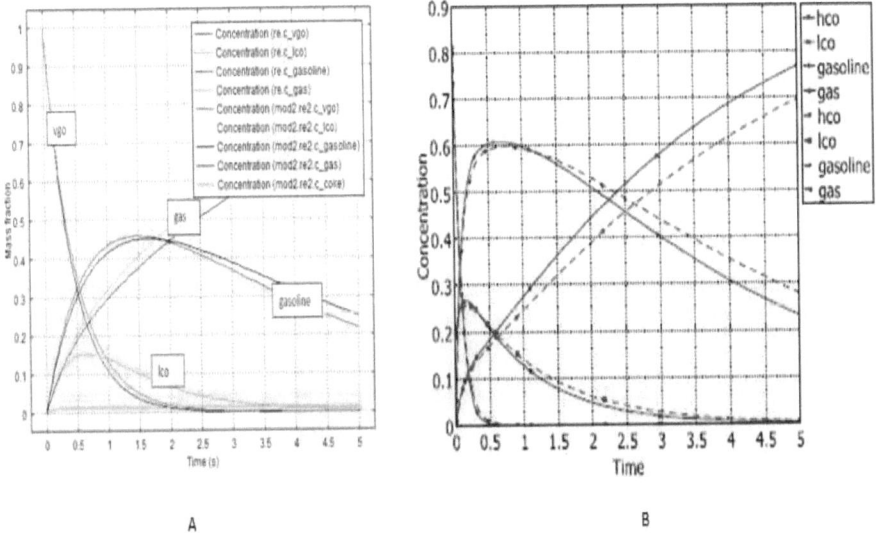

Figure 4.3: Comparison of model predictions with predictions from literature (A: Model predictions. B: Predictions from literature (Den Hollander et al., 2003))

It is also obvious from the results obtained (Figure 4.1 and Figure 4.2) that mass transfer resistance significantly affects FCC reactions. Generally, the curves stretch forward (in Figures 4.1 and 4.2), occupying a wider area in the graphs. By implication, the reactions do not occur (instantaneously) as predicted in literature (Figure 4.3). This is because the resistance to mass transport causes a significant delay in the movement of species from the gaseous phase (bulk gas) to the catalyst and within the catalyst to the active site of the catalyst where reaction occurs. The model result presented in Figure 4.2 shows that the overcracking of gasoline actually begins 1.5s after the commencement of the reactions. The gasoline overcracking point as predicted from literature however is 0.75s (Figure 4.3). This underprediction of the overcracking point could have adverse effects on reactor design and operation. Moreover, it can be observed in Figure 4.2 that the yields of gasoline and gas when there is coking of the catalyst as well as the yield of gasoline for no-coking condition are coincident at time, t = 2 seconds (which time corresponds to the riser's exit). The physical significance of the point of coincidence of the curves is that it defines the optimum residence time for risers provided that gasoline is the desired product. This explains why risers are designed to have a residence time of 2 seconds. However, the optimum riser residence time predicted from literature (Figure 4.3) is 2.5s which is greater than the optimum. Hence, when the two undesired phenomena (that is, resistance to mass transfer and catalyst coking) that are observed to occur in reality in fluid catalytic cracking were incorporated in the riser model presented here, the model predictions became more accurate.

In Figure 4.2, the effect of coking on the conversion/yield and selectivity is vividly depicted. The first pair of curves for the reactant (VGO) shows the effect of coking on the rate of conversion of the reactant. Comparing Figure 4.2 with Figure 4.1 (model without coking but with mass transfer resistance), it is seen that the blue profile for VGO (Figure 4.1) shifts conspicuously (within the first two seconds of the reaction) to the right (the purple profile in Figure 4.2) as a result of coking of the catalyst. By implication, the conversion of VGO at any given point in time when there is coking of the catalyst is less than the conversion at the same time when there is no coking. For instance, from Figure 4.2, when there is no coking of catalyst, the feedstock conversion 1s after the commencement of the reactions is 90% (the blue curve). However, when there is coking of the catalyst, the conversion of VGO at the same time is 86%. The reason for the observed reduction in the conversion of the reactant is not farfetched. Catalyst coking reduces the activity of FCC catalyst, thereby increasing the time required for the conversion of the reactant.

In Figure 4.1, the green curve is the concentration profile for LCO when coking does not occur. The curve shifts upwards to the position of the yellow curve in Figure 4.2 as a result of coking. The reason for the shift being also that less LCO is converted per time when there is catalyst coking than when coking does not occur (Figure 4.1).

In Figure 4.2, the gasoline pair is represented by the red and black curves. The yield of gasoline product when catalyst coking occurs (the black curve) is below the profile of gasoline when there is no coking of the catalyst (the red curve in Figures 4.1 and 4.2) within the first two seconds of the reaction. This implies that coking reduces the yield of gasoline as a result of the decrease in the rate of conversion of VGO.

In Figure 4.2, the sky blue and the deep blue curves are for the gas lump. The sky blue curve (concentration profile of gas in Figure 4.1) moves down to the position of the deep blue curve (Figure 4.2) when coking of the catalyst occurs. This implies that there is a decrease in the yield of gas due to the deactivation of the catalyst by coke. Hence, coking also has a negative effect on the yield of gas in FCC reactions.

The plant data presented in Tables 4.1 and 4.2 show that the model results in this investigation are closer to reality than the results from previous investigators. The model parameter values that were used to solve the model were effective diffusivity, D_e: 5.871×10^{-9} m/s², particle overall effectiveness factor, η_0: 0.980, Thiele modulus, φ: 0.540 and mass transfer coefficient, k_g: 0.166 m/s. The improvement obtained in this present work is undoubtedly attributed to the inclusion of mass transfer effects into the modeling.

Table 4.1: Kaduna Refinery and Petrochemicals (KRPC) Fluid Catalytic Cracking Riser (FCCR) flow rates (Abridged version, May, 2013)

Flow, kg/h	\multicolumn{9}{c}{DAY}
	3	6	7	8	...	24	27	28	30	31	Average	% Yield
Feed	83.00	83.00	83.00	83.00	....	80.00	73.00	73.00	73.00	73.00	78.98	
Gasoline	20.00	23.00	50.00	37.00	...	40.00	25.00	38.00	33.00	34.66	43.94	
LCO	15.00	21.50	24.00	12.00	....	13.00	16.00	9.00	10.00	15.66	19.85	
DCO	8.00	11.70	12.00	10.00	....	20.00	16.00	12.00	14.00	13.52	17.14	
											% Feed conversion: 80.92	

Table 4.2: Kaduna Refinery and Petrochemicals (KRPC) Fluid Catalytic Cracking Riser (FCCR) design data

Species	Yield, wt%
Gas	17.88
Gasoline	50.00
LCO	15.15
Coke	5.08

Figure 4.4 shows that the activity of the catalyst drops abruptly from 1 (one) to an asymptotic value of 0.89 (within the first 0.5s of the reaction). This period (the first 0.5s of the reaction) actually predicts the duration of aggressiveness of the equilibrium catalyst. At the end of this period, the activity of the catalyst (a) becomes steady as depicted in Figure 4.4. The asymptotic value of catalyst deactivation predicted by the model in this work compares well with the prediction from literature (0.85) in Figure 2.7 (Den Hollander et al., 2003).

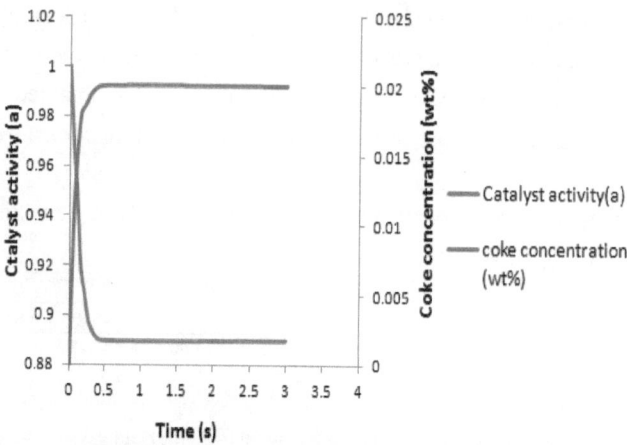

Figure 4.4: Coke concentration and catalyst activity decline in Fluid Catalytic Cracking (FCC) of Vacuum Gas Oil (VGO)

4.2 Reactor Model Results (MATLAB)

The results arrived at for the reactor geometry simulations were as presented in Figures 4.5 to 4.7.

The predicted yields of LCO, gasoline, gas and coke as depicted in Figure 4.5 are 14.50wt%, 47.86wt%, 17.11wt% and 4.89wt% respectively. These values compare favourably well with plant design data (Table 4.2). The maximum deviation of the predicted yield values from plant design data is 4.29% which is less than the error limit of 5%. Hence, the model here presented accurately mimics an industrial riser.

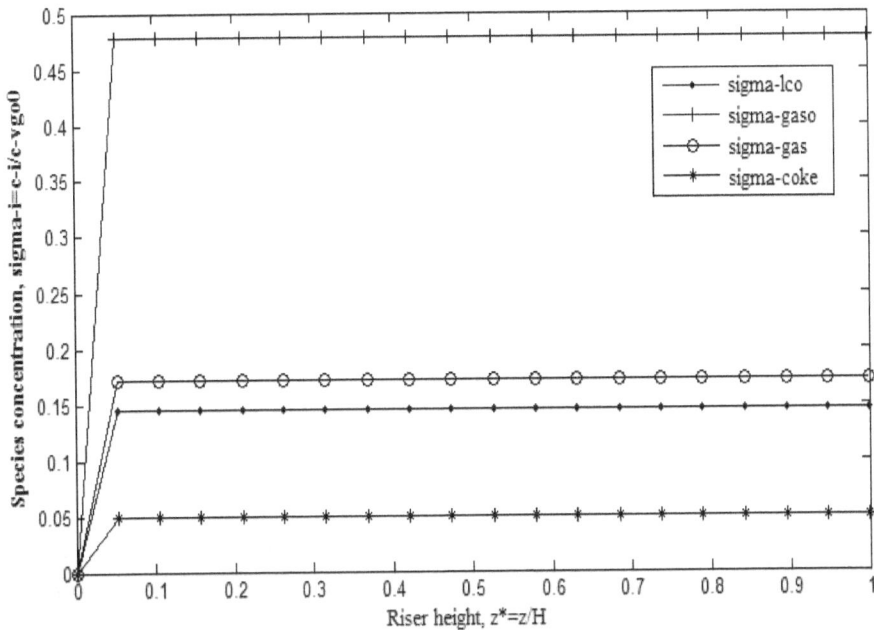

Figure 4.5: Products yield along riser height

Figure 4.6 presents the predicted conversion of VGO as a function of reactor height. A conversion of 80.19% was predicted by the model. (KRPC FCC riser was actually designed to operate at two feedstock conversion levels: 80% and 60%). The temperature profile along the height of the reactor is as given in Figure 4.7. The model result in Figure 4.7 predicts a reactor temperature drop of 38.99°C (a drop from 791.00K – 752.01K). The predicted temperature drop falls within the range predicted by previous researchers for industrial risers; 30-40°C (Gupta, 2006). In reality, VGO conversion and riser temperature drop both occur within the first 2-4m of the reactor. This is confirmed by Figures 4.6 and 4.7. VGO conversion and the riser temperature drop occurred in the reaction zone of the reactor as shown in Figures 4.6 and 4.7 because it is in the reaction zone of the riser that heat is absorbed by the feedstock from the incoming catalyst to vaporize the feedstock as well as initiate the endothermic cracking reactions. Hence, the reactor temperature is observed to drop in the same zone of the riser where feedstock conversion occurs (Figures 4.7).

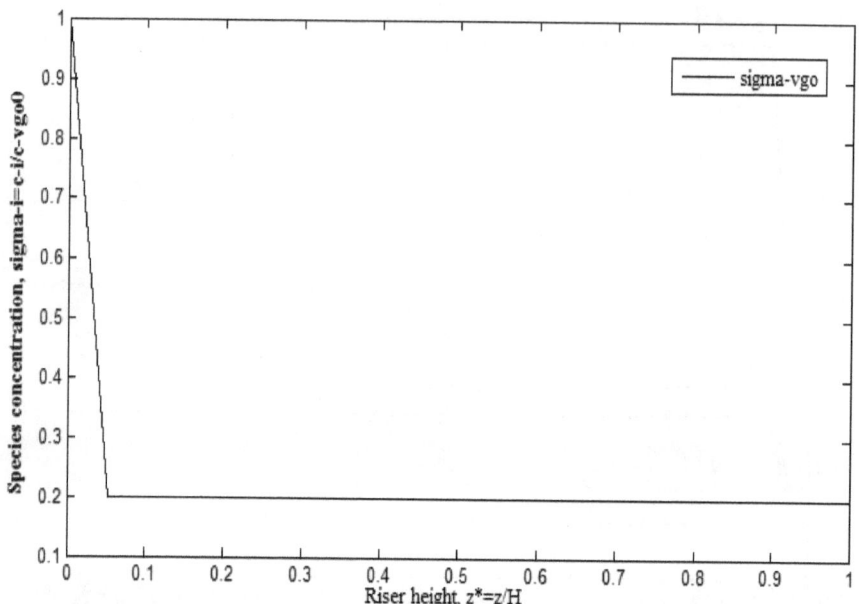

Figure 4.6: Conversion of Vacuum Gas Oil (VGO) along riser height

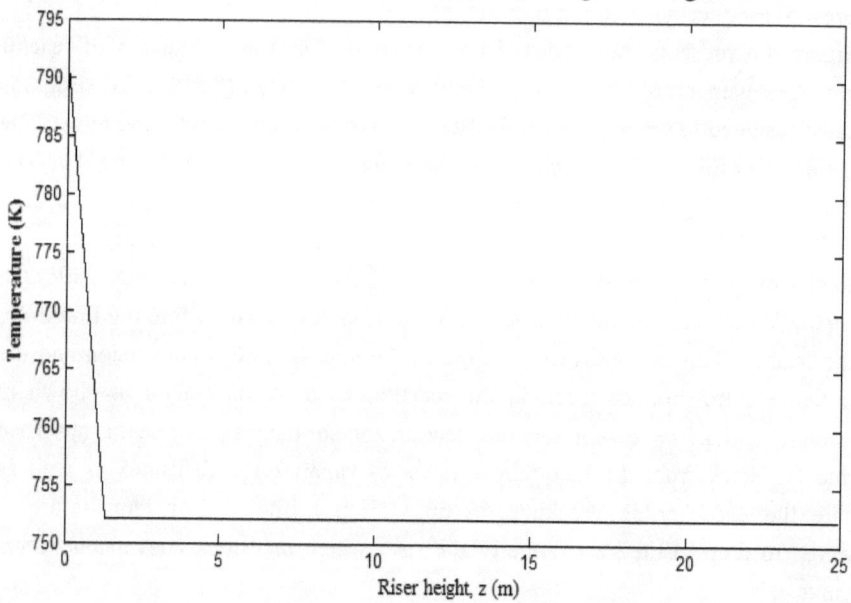

Figure 4.7: Temperature drop along riser height

The predicted conversion of vacuum gas oil as well as the temperature drop occurred within the first 2m of the riser length. This corresponds to the reaction zone of the riser as shown in Figure 4.8.

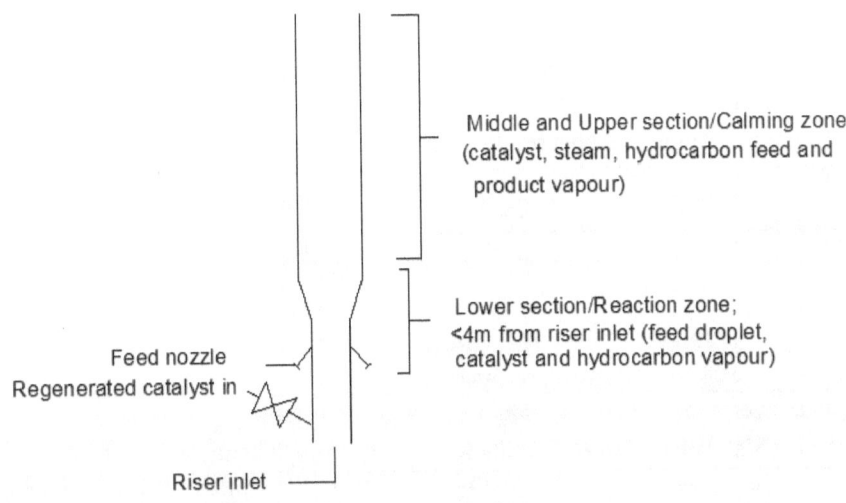

Figure 4.8: Sections of Fluid Catalytic Cracking (FCC) riser

The reason that the cracking reactions occur within the first 2-4m (the reaction zone) of the reactor is not farfetched. In the reaction zone of the riser, radial dispersion is highly pronounced as feedstock molecules move in the radial direction from the bulk gas phase into the catalyst to be cracked. Thereafter, there is expansion of product gases and the gases flow upward. Though the conversion of VGO and the temperature drop occur within the first 2-4m of the riser as shown in Figures 4.5 to 4.7, industrial risers are 20-35m in height. The extraneous height is to allow for flow development because flow meters and temperature sensors are usually located at the middle and at the top of the reactor where the flow is fairly or fully developed. Another reason that the height of risers is 20-35m is to allow for flexibility of operation when the desired product changes. If maximum gas production is the objective, then overcracking of gasoline becomes desirable. The middle section of the reactor (Figure 4.8) is the region of gasoline overcracking to maximize the yield of gas.

4.2.1 Validation of model results

The validation of model results with plant data and the results obtained from literature is as summarized in Table 4.3 and Table 4.4. Table 4.3 shows the feedstock conversion level as well as product yield as obtained from plant design data (PDD) and current plant operation data (CPOD). The first row of Table 4.3 gives the VGO conversion level while the fourth column of the table are the values of VGO conversion and product yields as predicted by the model.

Table 4.3: Validation of model results with plant data

Species	Conv./Yield wt% (PDD)	Conv./Yield wt% (CPOD)*	Conv./Yield wt% (Model)	% Dev. from design
VGO	80	80.92	80.19	0.24
LCO	15.15	19.85	14.50	4.29
Gasoline	50	43.93	47.86	4.28
Gas	17.88	Not available	17.11	4.31
Coke	5.08	Not available	4.89	3.74

* Values averaged over one (1) month steady plant operation

Table 4.4: Validation of model results with results from literature

Species	Conv./Yield, wt% (PDD)	Conv./Yield, wt% (Fernandes et al., 2003)	Conv./Yield, wt% (Ahari et al., 2008)	Conv./Yield, wt% (In this work)
VGO	80.00	78.00	78.00	80.19
LCO	15.15	10.00	-	14.50
Gasoline	50.00	48.00	45.00	47.86
Gas	17.88	18.00	-	17.11
Coke	5.08	5.00	5.00	4.89

Key:

Conv.: Conversion

PDD: Plant Design Data

CPOD: Current Plant Operation Data

Dev.: Deviation

VGO: Vacuum Gas Oil

LCO: Light Cycle Oil

Table 4.3 and Table 4.4 show that the model results compare favourably well with the data obtained from the plant as well as the results of previous researchers. The maximum deviation of the model predictions from plant design data was found to be 4.29% (< 5%). Hence, the error in the predictions falls within acceptable limit.

4.3 Simulation of Coking

The validated riser model was used to simulate coking in FCC unit riser. The coke yield and the catalyst-oil-ratio were predicted as a function of reactor temperature by varying the inlet temperature within a chosen range; 779K-791K (506°C-518°C). The result of the simulation is as displayed in Figure 4.9.

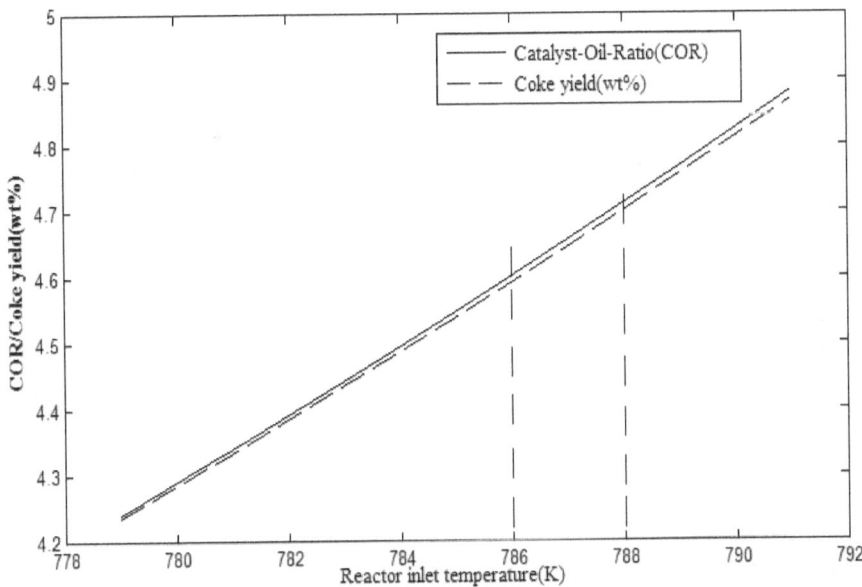

Figure 4.9: Model result for simulation of coking in Fluid Catalytic Cracking (FCC) riser

The plot in Figure 4.9 confirms that coke yield and reactor temperature both increase monotonously as the catalyst-oil-ratio increases. In existing plants, the FCCU reactor temperature is categorized into three regimes. The low operating temperature regime is the temperature range which quenches the reactions in the riser if the reactor is operated within the range. The high operating temperature regime on the other hand causes excessive coking of the catalyst at the expense of gasoline production. The desired operating temperature regime (the optimum temperature range) however, is the temperature range between the two extremes stated above. Based on the observed temperature regimes of industrial risers, Figure 4.9 could be categorized into three temperature regimes as follows:

i. Low operating temperature regime (T<786K): If the riser is operated in this regime (lower region of the graph where the curves taper towards each other), the reactions will quench. Hence, operation in this regime is not advisable.

ii. Optimal operating temperature regime (786K<T<788K): In this temperature range, COR and catalyst coke content profiles show no tendency of intersecting as shown in Figure 4.9. This is the regime of optimal riser operation (without excessive coking). Operating the FCC unit at temperatures that lower the rate of coking would reduce the cost of regeneration of the catalyst by reducing the volume of air that is needed for regeneration.

iii. High operating temperature regime (T>788K): In this temperature zone, the separation between the two curves is more pronounced than in the regimes of lower temperature. Unit operation in this temperature range is also not advisable because it leads to excessive coking and gas production at the expense of the most economical product (gasoline).

Figure 4.10 depicts the variation of COR and the yield of gasoline with reactor temperature. The average yield of gasoline within the proposed optimum temperature range (786K<T<788K) is 45.58%. This value falls within the range of gasoline yield predicted in literature and obtainable from existing plants (45%-50%).

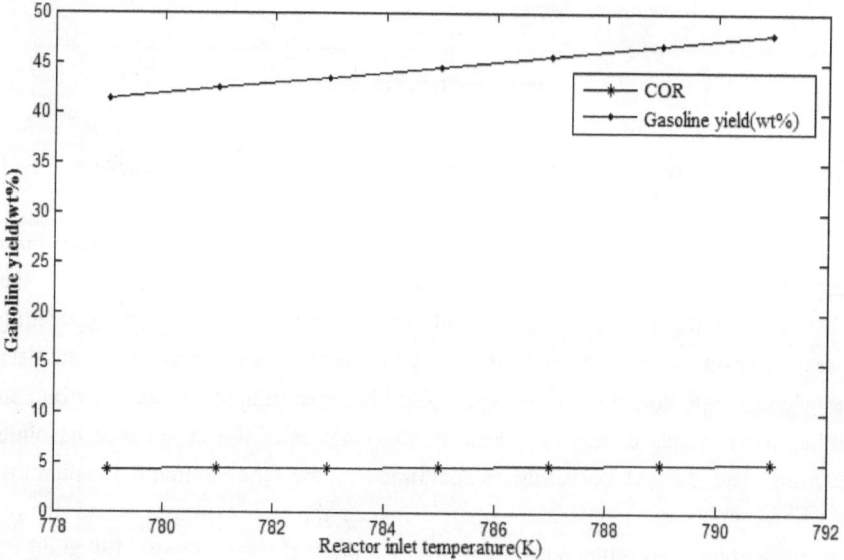

Figure 4.10: Variation of Catalyst-Oil-Ratio (COR) and gasoline yield with reaction temperature

CHAPTER 5
CONLUSIONS AND RECOMMENDATIONS
5.1 Conclusions
The following conclusions could be drawn from this work:

1. The simulation carried out in this project shows that mass transfer resistance actually plays a significant role in FCC reactions because it increases the precision of model predictions from 89.46% (for predictions from literature) to 96.63%. Hence, mass transfer resistance should not be neglected in the mathematical modeling of FCC reactions and the reactor.

2. The results of this investigation also show that an operating inlet temperature range of 786K<T<788K (513^0C<T<515^0C) and a catalyst-to-oil ratio (COR) range of 4.61-4.72 are optimal for fluid catalytic cracking of VGO.

3. Amongst the operational objectives of FCC units are plant operation at reduced coking rate (coke as a by-product is expensive to burn off the catalyst) and that gas production should not be in excess (gas is less valuable and yet expensive to compress). Plant operation within the optimal temperature range ensures that these economic objectives are achieved.

4. The predicted yield of gasoline by the model here presented is 47.86% with a VGO conversion of 80.19%, a coke yield of 4.89% and riser temperature drop of 38.99^0C (the degree of accuracy of the model predictions being 96.63%). The model predicted a riser residence time of 2s (the optimum residence time for which industrial risers are designed).

5. The model results obtained in this work compare favourably well with plant design data (50% gasoline, 80% VGO conversion, 5% coke and 30-40°C temperature drop). Hence, the FCC riser model presented here accurately mimics the FCC riser of KRPC Ltd.

6. The softwares employed for this investigation (COMSOL Multiphysics 4.0 and MATLAB) have also been able to model and simulate an industrial Fluid Catalytic Cracking (FCC) riser.

Contributions to knowledge

This research work has contributed to the existing knowledge on fluid catalytic cracking by:

a) Accurately predicting the conversion of Vacuum Gas Oil (VGO), the yield of the major species in fluid catalytic cracking of VGO, and the temperature drop of the reactor in fluid catalytic cracking of VGO.

b) Accurately predicting the residence time of industrial FCCU risers (2s; the optimum residence time for which industrial risers are designed).

c) Proving that mass transfer resistance actually plays a significant role in FCC reactions and as such it should not be neglected in the modeling of FCCU risers.

d) Depicting how the optimal temperature range for the riser reactor could be predicted from the simulation of coke that is deposited on FCC catalyst during the cracking reactions.

5.2 Recommendations

The major challenge encountered in carrying out this work was acquisition of reliable plant data for running the model as well as for model validation. This problem was further compounded by erratic plant operation which made the online data that was obtained from the plant to be unreliable. This challenge can be overcome by the development of prototype risers or riser simulators.

REFERENCES

Ahari, J.S., Farshi, A. and Forsat, K. (2008). A Mathematical Modeling of the Riser Reactor in Industrial FCC Unit. *Petroleum and Coal* 50(2), 15-24. Retrieved July 24, 2012 from www.vurup.sk/sites/...sk/.../PC_2_2008_Sadeghzadeh.pdf.

Ali, H., and Rohanni, S. (1997). Dynamic modeling and simulation of a riser-type fluid catalytic cracking unit. *Chem. Eng. Technol.*, 20, 118-130.

Alsabei, R.M., (2011). Model Based Approach for the Plant-wide Economic Control of Fluid Catalytic Cracking Unit. pp. 12.

Bessiris, Y., and Harismiadis, V. (2007). *Dynamic Simulation of a Fluid Catalytic Cracking Unit*. Paper presented at 10th Topical Conference on Refinery Processing- TA007 Advances in FCC. Pune, India. pp. 1.

Chen, L., and Weinstein, H. (1993). Shape and Extent of the Void formed by a Horizontal Jet in a Fluidized Bed. *AICHE J.*, 39, 1901-1909.

Corella, J., Bilbare, R., Molina, J.A., and Artigas, A. (1985). Variation with Time of the Mechanism, Observable Order and Activation Energy of the Catalyst Deactivation by Coke in the FCC Process. *Ind. Eng. Chem. Process Des. Dev.*, 24, 625-636.

Corma, A., and Martinez-Triguero, J. (1994). Kintics of Gas Oil Cracking and Catalyst Decay on SAPO-37 and USY Molecular Sieves", *App Catal*, 118, pp.153-162.

Coxon, P.G., Bischoff, K.B. (1987). Lumping Strategy. Introduction to Techniques and Application of Cluster Analysis. *Ind. Eng. Chem. Res.*, 26, 1239-1248.

Den Hollander, M.A., Makkee, M., and Moulijn, J.A. (2001). Prediction of the Performance of Coked and Regenerated Fluid Catalytic Cracking Mixtures. Opportunities for Process Flexibility. *Ind. Eng. Chem. Res.*, 40, 1602-1607.

Den Hollander, M.A., Wissink, M., Makkee, M., and Moulijn, J.A. (2003). Fluid Catalytic Cracking. *J. Appl. Catal. A*, 223, p103.

Fernandes, J.L., Pinheiro, C.I.C., Oliveira, N. and Bibeiro, F.R. (2003). Modeling and Simulation of an Operating Industrial Fluidized Catalytic Cracking (FCC) Riser. pp. 1-4.

Forissier, M. and Bernard, J.R. (1991). Deactivation of Cracking Catalyst with Vacuum Gas Oil, In Catalyst Deactivation. Edited by Bartholomew, C.H., and Butt, J.B., pp359-366.

Forissier, M. Formenti, M.,and Bernard, J.R. (1991). Effect of Total Pressure on Catalytic Cracking Reaction. *Catal Today*, 11, pp73-83.

Froment, G.F., and Bischoff, K.B. (1979). Chemical Reactor Analysis and Design. New York: John Wiley.

Geankoplis C.J. (2011). *Transport Processes and Separation Process Principles*. (4th ed.). New Jersey, U.S.A. Pearson Education Inc., 2011. pp. 425, 482.

Gerber, M.A., Frye, J.G., Bowman, L.E., Fulton, J.L., Silva, L.J. and Wai, C.M. (1999). Regeneration of Hydrotreating and FCC Catalysts. Retrieved from www.pnl.gov/main/publications/external/.../PNNL-13025.pdf

Gupta, R.S. (2006). Modeling and Simulation of Fluid Catalytic Cracking Unit. Retrieved June 12, 2012 from http://www.dspacethara.edu:8080/dspace/bitstream/123456789/56/3/T56.pdf.

Heydari, M., Ebrahim, H.A. and Dabir, B. (2010). Modeling of an Industrial Riser in the Fluid Catalytic Cracking Unit. Retrieved November 7, 2013 from www.core.kmi.open.ac.uk/downloadpdf/5575427

Jacob, S.H., Gross, B., Voltz, S.E., and Weekman, V.W., Jr. (1976). A Lumping and Reaction Scheme for Catalytic Cracking. *AIChE J.*, 22, 701-713.

Juarez, J., Lopez-Isunza, F., Aguilar-Rodriguez, E., and Moreno-Mayorga, J.C. (1997). A Strategy for Kinetic Parameter Estimation in the Fluid Catalytic Cracking Process. *Ind. Eng. Chem. Res.*, 36(12): 5170-5174.

Kraemer, D., Larocca, M., and de Lasa, H.I. (1991). Deactivation of Cracking Catalyst in Short Contact Time Reactors: Alternative Models. Can. *J. Chem. Engr.*, 69, 355-360.

Kraemer, D.W. and de Lasa, H.I. (1988). Catalytic cracking of hydrocarbons in a riser simulator. *Ind. Eng. Chem. Res.*, 27 (11), 2002-2008.

Kraemer, D.W., Larocca, M., and de Lasa, H.I. (1990). Deactivation of Cracking Catalysts in Short Contact Time Reactors: Alternative Models. Can. *J. Chem. Eng.*

Larocca, M., Ng, S., and de Lasa, H. (1990). Catalytic Cracking of Heavy Gas Oils: Modeling Coke Deactivation. *Ind. Eng. Chem. Res.*, 29(2), 171-180.

Lee, L., Chen, Y., Huang, T., and Pan, W. (1989). Four-lump Kinetic Model for Fluid Catalytic Cracking Process. Can. *J. Chem. Eng.*, 67, 615-619.

Lopes, G.C., Rosa, L.M., Mori, M., Nunhez, J.R. and Martignoni, W.P. (2012). CFD Study of Industrial FCC Risers: The Effect of Outlet Configurations on Hydrodynamics and Reactions. Retrieved November 7, 2013 from http://www.hindawi.com/journals/ijce/2012/193639/

Lumping and Modeling of FCC Reactions. (1996). Retrieved August 22, 2012, from faculty.kfupm.edu.sa. pp. 7-10.

Mao, X., Weng, H., Zhu, Z., Wang, S., and Zhu, K. (1985). Investigation of the Lumped Kinetic Model for Catalytic Cracking III. Analyzing Light Oil Feed and Products and Measurement of Kinetic Constants. Acta Pet. Sin. (Pet. Process Sect.), I, II.

McFarlane, R.C., Reineman, R.C., Bartee, J.F. and Geogarkis, C. (1990). Dynamic Simulator for a Model IV Fluid Catalytic Cracking Unit. Retrieved May 25, 2012 from www.acslx.com/support/techpapers/chemical/246pdf.

Merry, J.M.D. (1971). Penetration of a Horizontal Gas Jet into Fluidization Bed. *Trans. Inst. Chem. Eng.*, 49, 189-195.

Missen, R.W., Mims, C.A., and Saville, B.A. (1999). *Introduction to Chemical Reaction Engineering and Kinetics*. New York. John Wiley & Sons Inc., 1999. Pp 198-214, 524-527.

National Agency for Science and Engineering Infrastructure. (2010). *Introductory Training on Advanced Manufacturing Technology*. pp. 61.

Oliveira, L.L. (1987). Estimacao de Parametros e Avaliacao de Modelos de Craqueamento Catalytico. M.Sc. Thesis (in Portuguese). University Federal do Rio de Janeiro, Brazil.

Pareek, V.K., Adeshina, A.A., Srivastava, A., and Sharma, R. (2003). Modeling of a Non-isothermal FCC Riser. *Chem. Eng. J.*, 92, 101-109.

Praveen, C.H., and Shishir, S. (2009). Effect of pressure on height of regenerator dense bed in an FCCU. *Petroleum & Coal* 51 (2) 124-135.

Pitault, I., Nevicato, D., Forissier, M., and Bernard, J.R. (1994). Kinetic Model on a Molecular Description for Catalytic Cracking of Vacuum Gas Oil. *Chem. Eng. Sci.*, 49, 4249-4262.

Kellogg, P. (1980). *NNPC Kaduna Refinery Project Operating Manual for Process Unit FCC* (Vol. VIII-1, pp. 1, 3, 20-22, 32). Yokohoma, Japan: Chiyoda.

Sa, Y., Chen, X., Liu, J., Weng, H., Zhu, Z., and Mao, X. (1985). Investigation of the Lumped Kinetic Model for Catalytic Cracking and Establishment of the Physical Model. Acta Pet. Sin. (Pet. Process sect.), 1, 3.

Sa, Y., Liang, X., Chen, X., and Liu, J. (1995). Study of 13-Lump Kinetic Model foe Residual Catalytic Cracking. Selective papers in memorial of 30th anniversary of fluid catalytic cracking process in China, Luoyan, Petrochemical Engineering Corporation, Luoyang, China, p. 145.

Simonsen, B. (2008). Heat Exchange in a Fluidized Bed Calcination Reactor. Retrieved from https://daim.idi.ntnu.no/masteroppgaver/IME/.../masteroppgave.pdf pp. 34

Souza, J.A., Vargas, J.V.C., Von Meien, O.F. and Martignoni, W.P. (2007). Modeling and Simulation of Industrial FCC Risers. Retrieved November 7, 2013 from http://deniec.ufpr.br/reterm/ed_ant/11/artigo/technol01.pdf

Speight, J.G. (2007). *The Chemistry and Technology of Petroleum.* (4th ed.). London, CRC press, Taylor and Francis Group, 2007. pp. 556-557, 571, 573, 576-579.

Szepe, S. and Levenspiel, O. (1971). Catalyst Deactivation. Chemical Reaction Engineering, Proceeding of the Fourth European Symp., Brussels, Pergamon Press, Oxford, 265.

Takatsuka, T., Sato, S., Morimoto, Y., and Hashimoto, H. (1987). A Reaction Model for Fluidized-bed Catalytic Cracking of Residual Oil. *Int. Chem. Eng.*, 27(1), 107-116.

Weekman, V.W.,Jr. (1968). A Model of Catalytic Cracking Conversion in Fixed, Moving and Fluid-bed Reactors. *Ind. Eng. Chem. Process Des. Dev.*, 7, 90-95.

Weekman, V.W.,Jr., and Nace, D.M. (1970). Kinetics of Catalytic Cracking Selectivity in Fixed, Moving, and Fluid-bed Reactors. *AIChE J.*, 16, 397-405.

Yen, L., Wrench, R., and Ong, A. (1987). *Reaction Kinetic Correlation for Predicting Coke Yield in Fluid Catalytic Cracking.* Paper presented at the Katalistisks' 8th Annual Fluid Catalytic Cracking Symposium, Budapest, Hungary.

Yen, L.C., Wrench, R.E., and Ong, A.S. (1988). Reaction Kinetic Correlation Equation Predicts Fluid Catalytic Cracking Coke Yields. *Oil Gas J.*, 86, 67-70.

You, J. and Zhu, C. (2008). Hydrodynamic model of Fluid Catalytic Cracking (FCC) Riser Reactor. Retrieved November 7, 2013 from www.nt.ntnu.no/users/skoge/prost/proceedings/aiche_2008/data/papers/p130616.pdf

Zhu, K., Mao, X., Weng, H., Zhu, Z., and Liu, F. (1985). Investigation of the Lumped Kinetic Model for Catalytic Cracking: II. A prior simulation for experimental planning. Acta Pet. Sin. (Pet. Process Sect.), 1, 47.

I want morebooks!

Buy your books fast and straightforward online - at one of the world's fastest growing online book stores! Environmentally sound due to Print-on-Demand technologies.

Buy your books online at
www.get-morebooks.com

Kaufen Sie Ihre Bücher schnell und unkompliziert online – auf einer der am schnellsten wachsenden Buchhandelsplattformen weltweit!
Dank Print-On-Demand umwelt- und ressourcenschonend produziert.

Bücher schneller online kaufen
www.morebooks.de

OmniScriptum Marketing DEU GmbH
Heinrich-Böcking-Str. 6-8
D - 66121 Saarbrücken
Telefax: +49 681 93 81 567-9

info@omniscriptum.com
www.omniscriptum.com

www.ingramcontent.com/pod-product-compliance
Lightning Source LLC
Chambersburg PA
CBHW031538210526
45464CB00003B/1063